Time
in French life and thought

Time
in French life and thought

by Richard Glasser
translated by C. G. Pearson

Manchester University Press
Rowman & Littlefield, Totowa, N.J.

Studien zur Geschichte des französischen Zeitbegriffs
first published 1936 by Max Hueber Verlag, Munich

Published by the University of Manchester at
THE UNIVERSITY PRESS
316–324 Oxford Road, Manchester M13 9NR

UK ISBN 0 7190 0464 0

USA
Rowman & Littlefield, Totowa, N.J.
US ISBN 0 87471 077 4

Made and printed in Great Britain by
Butler & Tanner Ltd, Frome and London

Contents

Introduction

The realization that the concept of time formed by an individual or by a stage in intellectual history, and the manner in which this is treated, tell one something essential concerning the individual or culture in general is nothing new. Great minds, or intellectual currents, conscious of the innovations desired by them, in earlier centuries interpreted the new positions they adopted in their world view or conduct of life, in which they set themselves against traditional elements, as antithetical concepts of time also. Rabelais, Rousseau and the Romantics are cases in point. For them the clock was an intellectual crossroads. They were aware of the basic historical fact that man's attitude to time is intimately bound up with his spiritual make-up and with his general behaviour, in both its active and passive aspects. One's attitude towards, and evaluation of, time is not an independent entity, nor is it subordinate to other elements in one's consciousness, but everywhere and at all times informs them. This circumstance was, indeed, always considered as axiomatic and hence was never formulated as a general judgment, but it was always taken into account. What Rousseau said against society's conception of time is not to be isolated from his criticism, generally considered, of society, but in itself constitutes this criticism.

It was not until much later, however, that that branch of science which makes man in his historical context the object of its study proceeded methodically to appraise the importance of the concept of time from the historical point of view and systematically to investigate changes in the awareness of time in bygone epochs. Only in recent decades has science noticeably concerned itself with the question of man's changing relationship to time. The French philosopher Henri Bergson played no small part in arousing interest in the subject. Although he himself bore no

relationship to history, yet, by virtue of his differentiation between an external, spatially and quantitatively conceived time
and an inner, unextended, irreversible and qualitatively experienced duration (*durée*), he afforded new modes of judgment to
research in the domain of intellectual history and thereby
imparted fresh stimuli and posed new problems to it. Oswald
Spengler, for example, appropriated the Bergsonian definition of
time in his well known book. What he says in *The Decline of the
Western World* on the nature of time betrays the influence
exerted on him by the French thinker. The literary critic Albert
Thibaudet underwent the same influence at the hands of the
author of *Les Données immédiates de la conscience*. In his essays,[1]
as also in his book *Le Bergsonisme*, Thibaudet utilized the concepts
of space and time in elucidating processes and situations in the
history of thought, with valuable results. But since he did so
incidentally and for purposes of guidance rather than consistently
and scientifically, he was inclined to emphasize the weight and
importance given by consciousness to space and time rather than
the particular form which imparts to an historical epoch its
conception of space and time.

Werner Gent has written a history of the concept of time (*Das
Problem der Zeit, eine historische und systematische Untersuchung*,
Frankfurt, 1934) in the historical section of which he sets out the
development of his philosophical doctrines on the subject of time;
this development, however, is outside the scope of our inquiry.

It must be pointed out that the views of individual philosophers
or even of entire philosophical movements on the question of time
do not form the focal point of our considerations. Our historical
investigation of changes in the French conception of time does not
aim at tracing modifications in the philosophical apprehension of
time. Time is not simply an object on which the power of human
understanding is directed, but an element which must be
informed. We wish to demonstrate its significance for the general
consciousness of a cultural era—that is to say, how it is judged, its
validity in relation to other values, its influence on human activity,
the place it is allotted as an informable element, its possible
significance in the domain of the beautiful and the good, what
attention it receives, and so on. All this evidences itself with
greater clarity in the more concrete manifestations of life and

culture as they are represented in such spheres as literature, the exercise of moral values, the retrospective view of history, public life, jurisprudence, social intercourse, the shaping of an individual life and so forth, than in philosophical apprehensions and constructions. Whereas man can evade philosophizing on the subject of time, he is compelled by life to come to terms with it and to decide what he is to make of it. A definite attitude on the part of humanity *vis-à-vis* time is inherent in every cultural manifestation. This applies also to that all-embracing manifestation of human culture, language.

If, in this work, we wish to treat of the history of the various apprehensions of time largely, though not exclusively, from a linguistic standpoint, we find ourselves confronted with the problem of the relationship between time and language. Does language adequately express the varying attitudes of humanity towards time? Can a purely inward experience be reproduced in all its inwardness in language? Bergson's view of the nature and function of language obliges him to answer this question in the negative. According to him, a purely inward awareness of duration, consisting of a sequence of interpenetrating heterogeneous instants, not numerically apprehensible, cannot find adequate expression in language. Every psychic event is something that occurs only once, and therefore every experience, by virtue of its psychic uniqueness, is unrepeatable. Language abstracts itself from such uniqueness, Bergson thinks, because it enumerates inner experiences and distinguishes them from one another. Events that are indefinable, because they are not spatially separable, are expressed by means of words, which represent things. Qualitative variety is falsified by words: 'Brutal words, with their well defined contours, words, which are the storehouse of stable, shareable, and consequently impersonal elements in the impressions received by humanity, extinguish or at least blur over the delicate and fleeting impressions received by our individual consciousness.'[2] This alleged incapacity of words to reproduce psychical experiences is at the heart of Bergson's view of the nature of language. For him, language is no more than a concept, a means of differentiation and delimitation, a species of social institution, the sole purpose of which is communication on the practical level.

If we view the matter from another standpoint we likewise

A*

become aware of this gulf between experience and language. Let us consider the manner in which we falsify by our language those temporal relations with which experience familiarizes us. Without any intention of being untruthful, we distort true temporal relations in such a way that we contract duration to a point and extend individual instants.

If we relate a simple circumstance, the temporal relations governing the individual elements that go to make it up are distorted by the length of time we take to recount them. In a sentence such as 'I rose, washed, breakfasted and then went walking for a few hours' brief acts such as getting up and washing are extended by the symbolism of language relatively to the act of going for a walk, which in reality took much longer. By dint of speaking and thinking in speech we continually and unconsciously lift events out of the stream of happening, assign their importance to them by the mere fact of naming them and attribute to them as we do so a length or brevity they do not possess. There is thus an unbridgeable gulf between reality and language, and Bergson appears to be right when he speaks (with the aid of another argument, it is true) of the impotence of language *vis-à-vis* time. It seems to be a result of man's position between the infinitely great and the infinitely small if he contracts extensions and stretches out points, in order to represent by means of language his own world lying between these two extremes. When he reports and represents circumstances, he inevitably emphasizes one period of time at the expense of another. Human language is not a ciné camera which indiscriminately reproduces successive moments of inner or external occurrences. To eliminate such undue emphasis in recounting circumstances and to reproduce an experience in the temporal relations corresponding to reality, without regard to their individual elements, would be the ideal of a time-oriented system of language. Marcel Proust attempted to realize this end in his work. In his book *A la Recherche du temps perdu* he strove to eliminate the dramatization of experience and the compression of events inherent in everything that is expressed in language, so that he might accurately recapture a period of time.

If true temporal relations are distorted by language to such an extent that a remarkable literary work, written with a definite purpose in mind, is required, not to realize the ideal of 'temporal orientation', but to show in what direction reality lies, it would appear that language and time are indeed irreconcilable and bear no essential relationship to one another. Such a relationship, however, is possible. Even if human language cannot succeed in representing true temporal relations and extensions by means of discourse of corresponding duration, it is nevertheless capable, within its own limits, of coming to terms with time. Our notion of duration enables us both to express external temporal extensions by means of concepts—when we speak of a second, a month, a century and so on—and to bring our consciousness to imprecise yet immediate awareness of time as an inner experience by means of obvious forms of expression. There are two modes of representing time, corresponding to the two modes of apprehending it: direct and indirect. Through the latter, time is registered and our knowledge of it enlarged. Through the former it is made accessible to our intuition as an impression and a process, it being the preserve of language to express duration, a function it fulfils with peculiar appropriateness. Since language is itself an entity in time, it is able to reproduce events with that property most essential to them, succession in time. A circumstance or a succession of circumstances may be symbolized in language, either successively or in isolation. The symbol is an experience no less than that which is symbolized. We are thus able, instead of differentiating between experience and language, to effect a *rapprochement* between experience of things and experience of language. Although both may lie on different planes, it is possible, all the same, to symbolize things by virtue of the intensity and the peculiar qualities of the experience of language.

Whereas speech, on the one hand, betrays by its conceptual qualities its remoteness from true time, whereby it is precluded from coming to terms with time as an experience by measuring it, it is nevertheless possible (despite Bergson, whose view of language is excessively narrow and who refuses to accord it its due importance) to reproduce experience of time in words not only faithfully but also in the form of melody, rhythm, expectancy and relief of tension. All these qualities and effects are common to both language and music, but whereas in the case of music tensions and stirrings of memory fulfil a purely formal function, language is able to bring into play its own conceptual and emotional associations, the reciprocity of question and answer. Language is a symbol of time no less than music.

According to whether one interprets time as proportion and extent or as an inner melody, one will prefer either the direct or the indirect means of its reproduction in language. Michel Bréal adumbrated the necessity of a history of the concept of time before he underwent the influence of Bergson and interpreted it as a linguistic problem when he investigated the forces determining the development of the verb. According to him, time was not one of the earliest influences to give rise to the existence of verbs. Time plays no part in the mind and language of primitive peoples. 'La notion claire du temps fait défaut aux populations restées à un état peu avancé de culture. Les voyageurs nous apprennent qu'au-dessous d'un certain degré de civilisation, il n'y a ni passé ni avenir.'[3] Nor do animals, naturally, have any notion of time in this sense. With the advance of culture the idea of time became clearer and more conceptual. Thus one readily comes to understand differences in the interpretation of time as a line of demarcation between a state of nature and one of civilization, between the educated classes and the masses, between the earlier and the later. As we review the history of French thought we shall see an ever greater conceptual clarity in the representation of temporal relations crystallize out in language as rationalism, culture and civilization progressed.

In his book *Frankreichs Kultur im Spiegel seiner Sprachentwicklung* Karl Vossler observed that changes in the interpretation of time were reflected in actual linguistic usage. He pointed to the use of tenses in Middle French to show how, in the Middle Ages, 'the subjective, non-spatio-temporal, but purely inward sense of time had become increasingly less current as an experience', and also how, in the sixteenth century, the subjective view of time took its place alongside the objective.[4]

Others, following in Vossler's footsteps, have observed a conscious relationship between a fact of linguistic history and the view of time prevalent in a particular epoch. In her essay 'Die uneigentlich direkte Rede' Gertraud Lerch noted how the introverted conception of time in the Renaissance was reflected in the increasing use of pseudo-direct speech in the sixteenth century.[5]

An attempted history of the concept of time as reflected in language need not confine itself to a particular grammatical domain. It is true that the findings of Vossler and Gertraud Lerch stemmed from a study of the verb, but they might equally well have had their source in a study of the vicissitudes of the noun or of the adverb. Moreover, changes in style, vocabulary and semantics, no less than changes in syntax, reflect modifications in the apprehension of time. The fact that the development and fate of a single word tell one something of the feeling for time that led to its formation is demonstrated by Leo Spitzer in his inquiry into the origin of the word *aujourd'hui*.[6] In its nature and development this word is evidence of a heightened sense of temporal order. Its function is not to reproduce an experience of time but to fix a particular point on the scale of temporal entities. Before this, Spitzer had established a relationship between the stylistic use of the word *Tag* (day) in various languages and psychological facts, when he pointed out that the succession of words *noctem et diem, nuit et jour* has an emotional and subjective basis.[7] A paper by A. Meillet, 'L'Emploi du génitif dans l'expression "de nuit" ', points even more strongly to a (tacit) relationship between a linguistic fact and a temporal concept.[8]

Studies such as these touched incidentally on the question of the concept of time, without investigating it more thoroughly, because it was irrelevant to the ends aimed at. An account, properly so called, of a particular culture's conception of time from a linguistic standpoint has so far been undertaken only by an historian: Hermann Fränkel, in his essay 'Die Zeitauffassung in der archaischen griechischen Literatur'.[9]

In this work an attempt is made to establish a close *rapprochement* between the development of the Greeks' conception of time and that of their general intellectual history. This must be the principal aim of anyone concerning himself with the history of the sense of time, for only if one is convinced that the attitude of an historical epoch *vis-à-vis* time must be viewed as an essential element of its culture is one entitled to view the attitude of humanity towards time in its historical perspective. In this matter it is a question—let us reiterate this—less of the part played by time in philosophical systems than of the manner in which time was *informed* by human volition, feeling, creativity,

life and points of view. This tacit principle is accepted by Fränkel
when he interprets individual words in the body of Greek poetry
as pointing to changes in the apprehension of time.

The most penetrating and pertinacious of such inquiries into
the concept of time in a particular epoch is the cardinal work of
Paul Imbs, *Les Propositions temporelles en ancien français. La
détermination du moment. Contribution à l'étude du temps grammatical français* (Paris, 1956). By means of this study, as de-
tailed as it is comprehensive, of syntactic–temporal forms of
expression, constantly changing as regards their function, meaning and importance, the author not only elucidates a lengthy and
many-sided linguistic process but also affords us an insight into an
awareness of time that differed from the modern view. Imbs sets
individual facts before us which combine to demonstrate how the
medieval linguistic community strove, with a persistence enduring throughout generations, towards greater clarity and order in
the representation and expression of temporal relations.[10]

Likewise, therefore, our researches are concerned chiefly,
though not exclusively, with facts of linguistic history when we
attempt to throw light on the question of the concept of time.
Developments in philological history are of importance for us
only in so far as they afford a guiding principle in the general
framework we desire to set up. In view of the fact that it aims to
provide, for the first time, a coherent conspectus of the French
view of time from the beginnings of literary history up to the
philosophy of Bergson, a work such as the present can claim to
afford only some measure of general guidance—guidance which
often requires amplification, clarification, amendment and
supporting matter. The reality of historical continuity and the
many guises in which it appears are advisedly neglected, not as
something inexistent, but as a remote goal. Thus if we speak of
the concept of time in the Renaissance, we do not do so in the
manner of a medieval realist who believes in the existence of the
general concept. We do so in the knowledge that we 'have re-
course to abstract concepts only to provide a convenient and
intelligible *aperçu* of a welter of diverse phenomena and
personalities'.[11]

For the latter-day person the awareness of time is something
highly individual. Everyone bears within him his personal

feeling for time, as much his own as his countenance. And yet a particular epoch has its own feeling for, and evaluation of, time, expressing its general intellectual tendency; the individual can hardly escape their influence. Participation in a great linguistic and cultural community—a participation which comprises, on the one hand, the process of development of temporal terminology and concepts, and, on the other, those elements informing life in general (elements which have much in common by virtue of their interrelatedness)—the act of drinking from the same cultural spring, the community of experience, memories and hopes born of involvement in the same point in historical process, a common historical panorama on which to look back; all these encourage the development of a collective feeling for time and life.

Notes

1 In *Nouvelle Revue Française.*
2 *Essai sur les Données immédiates de la conscience* (Presses Universitaires de France, 1946), ch. II, p. 98.
3 *Essai de Sémantique* (Paris, 1924), pp. 338–9.
4 *Op. cit.*, pp. 313–15; in the second edition, pp. 281–4. Such an occasional allusion to Bergson's distinction between *durée réelle* and *temps spatial* must not exceed the limits of admissible analogy. In his review of the first edition of our book (*Rom. Forsch.*, 50 (1936), pp. 329–36) Kurt Jäckel criticized the unmitigated application of Bergson's distinction to the time conception of former generations as recognizable in their linguistic manifestations. Without pleading for such an application we consider the distinction between measurable time and time as the experience of the flying hours a very fruitful one for investigations like the present.
5 'Idealistische Neuphilologie', *Festschrift für Karl Vossler* (Heidelberg, 1922), pp. 107f.
6 *Zeitschr. für rom. Phil.*, 50, 1926, pp. 336f.
7 *Aufsätze zur romanischen Syntax und Stilistik: 'noctem et diem'* (Halle, 1918), pp. 274–80.
8 *Mémoires de la Soc. de Linguistique de Paris*, 18, p. 240.
9 *Zeitschrift für Ästhetik und allgemeine Kunstwissenschaft*, 25, supplement, ed. Noack, pp. 97–118.
10 The interpretation of linguistic facts as revealing the time concept of a community throughout its social and cultural history forms only a part of the general development of philological thought,

which tends to look at linguistic history as a history of the human mind. Compare the close relationship established between language and the human concept of time by Fritz Tschirch, *Geschichte der deutschen Sprache* (Berlin, 1966), 1969, I, pp. 31–3, and II, pp. 33–5, 39–41.
11 Konrad Burdach, *Renaissance, Reformation und Humanismus* (Berlin, 1926), p. 101.

1

The concept of time in early medieval France

The absence of external time

The lives of the most primitive individuals and races are integrated with the rhythm of natural phenomena, either voluntarily or otherwise. Day and night and the course of the four seasons impose an elementary division of time on every life; this division is more or less a matter of necessity in the case of those peoples living in communion with nature, whereas in the civilized state of life man modifies this rhythm considerably, becomes independent of it and divests it of its elemental quality. The more one becomes aware of one's ability to direct one's life on a course departing from that dictated by nature, and the stronger are the bonds linking the human community, the more autonomously will one be able to shape one's own life in time. One's life thereby acquires a multitude of phases that overlap to a great extent, and of breaks which impart a measure of discontinuity to it. By this means human consciousness becomes increasingly able to survey phases in time, even when they lie outside one's own life. One gains the capacity to view historical periods, cultures, dynasties and political States in the place they occupy in time and one finally realizes one's own place in the history of humanity, or indeed of the world.

In the most ancient manifestations of the French mind, as they are represented in the national epic, this stage in the apprehension of time was far from being reached. It is true that in the external realities of the age of the Crusades one may observe some semblance of a coherent attitude towards time (in monastic life, for example), as also a standardized method of reckoning it; but the objective ordering of time played but a small part in the human imagination of that epoch, and in thought and feeling generally. Yet it would be unjust to call the individual of that age merely primitive, for the greater or lesser degree of culture attained by participation in intellectual developments within the ancient French linguistic community had its counterpart in certain levels of temporal orientation: '... du moins en milieu cultivé, l'expérience du temps avait largement dépassé le stade de pensée primitive'.[1] Time had not yet crystallized into an articulate image in human consciousness. Time elapsed, little attention being paid to its quantitative externalities: minutes, hours, weeks, months, and so forth. The need of, and capacity for, temporal orientation in the past and present was but small. One was not concerned with knowing one's place in time, for as yet there was no awareness of epochs. People were familiar only with the general and rather vague concepts of 'now' and 'formerly'.

The popular epics, which satisfied the historical needs of those centuries, were devoid of temporal content. The fact that roughly three centuries lay between the poet and the events he described was disregarded by the spirit of the *chanson de geste*. These works knew only an inward, emotional comprehensibility. They were narrated without any segmentation, one episode taking the place of the preceding one according to the necessities of internal linkage, and growing organically out of it. If, therefore, the *Chanson de Roland* was divided into episodes, as was done by Léon Gautier, Clédat and Eugen Lerch, this modernization fulfilled the practical purpose of affording guidance to the reader, but it was alien to the spirit of the national epic. The ability to represent a succession of external occurrences, despite the fact that they took place successively and were experienced as such, and to visualize them as spatially ordered, subdivided and related, was so far scarcely developed.

In the *chanson de geste* objective time was frequently distorted. In the *Chanson de Roland* Charlemagne is represented as a man two hundred years old.[2] In *Gaydon* he is spoken of as being actually even older.[3] In the *Chanson de Guillelme* Guillelme says of himself that he has a hundred and fifty years behind him.[4] There was no sense of temporal extension in narrative. There was no enumeration of years, no fixed temporal structure, no concern with the placing of personalities and events within that structure. The same held good in the case of the Greek epic: 'There was still no stable temporal framework to embrace epic events and apportion them to their due place'.[5] However, a comparison between the ancient French and the ancient Greek conceptions of time brings to light the important difference that, as early as the time of Homer, the latter was more abstract and philosophical than in the *chanson de geste*. Phrases such as 'the fateful day', 'the evil day', 'the pitiless day' do not occur in old French literature.[6]

A quality shared by both was that words with temporal connotations such as 'time' and 'day' had not yet acquired the wider significance we associate with them today. Χρονος, like the Old French *tens*, was used in a very specific and concrete sense. A further parallel lay in the path along which the concept of time was beginning gradually to move. 'In Homer we find an almost complete indifference to time, and at the start of the fifth century we find Chronos being almost overwhelmingly revered as the "father of all things".'[7]

This awareness of the significance of time was gradually gaining ground in French literature also from the eleventh to the fifteenth centuries. The part played by time in the life of man was unknown to the author of the *Chanson de Roland*. A circumstance to which a present-day poet might have attached a temporal significance had no such significance for him: Roland blows his horn, and Charlemagne hastens to the scene. How long can Roland go on fighting? How long does the Emperor need to hurry with assistance to his nephew's side? Will he arrive in time? These questions are not posed. In order to appreciate this absence of temporal sense it will be necessary for the reader to visualize all the techniques a modern writer would have brought to bear to create tension, so that he might bring home to us how every minute counted. This question of assistance materializing before it is too late is entirely disregarded by the epic poet. The Emperor, in fact, came on the scene too late by a matter of minutes; but these minutes are not recorded. It would have called for an objective eye for time.

The poet, however, experiences the struggle and his hero's fate as something outside a mere short span of time. Here, time is in fact something trivial in relation to the momentous events taking place. Roland's death would lose its quality of grandiose necessity if the idea of *malchance*, of an inept subordination to time, were bound up with it. This external time is of no significance for the fate of any man. Indications of time are not to be taken seriously. They tell one less about time than about the mentality of those who make them. Unconcerned with the possibilities of external time, which nevertheless have some measure of importance, the poet of the *Chanson de Roland* 'telescopes' events that are temporally widely separated. This is evidence, not of a conscious distortion of objective reality but of a conviction, peculiar to this century, of the unessential and unimportant quality of the measure of time. 'The Gothic poet could obliterate the limits of space and time as insignificant; he was able mentally to execute the wildest leaps and indeed to believe his audience likewise capable of them, which would have been impossible for anyone relying on his normal perceptions. Such a sensation produced in the audience, as in the modern reader, a feeling of abstraction from time and space, of a certain vacuity and unreality, similar to that engendered in the observer by a Gothic interior'.[8] We may therefore regard the interpretation of mundane events as expressed in the French national epic as 'timeless' if we read into this interpretation the mental attitude which developed the concept of time as a quantity and its projection into spatially visualized entities only slightly or not at all, attaching as it did no higher value to these matters. The author of *Raoul de Cambrai* offers certain vague indications of time:

> Une grant piece demora puis ensi,
> Mien esciant un an et quinze diz.   [vv. 805ff.]

Elsewhere in the same work we find the following:

> A Ribuemont fu Ibers li cortois,
> Et Berniers et Sain Quentin ses drois,
> Avuec sa femme qui molt l'anma en foi.
> Puis fu ainsis un an et quinze mois.   [vv. 6582ff.]

One readily sees what one is to make of such stereotyped turns as this. If, however, a serious attempt at indicating time is made, it is done with the purpose of conveying the duration of an event:

> *Guillelme:*
> Cele bataille durat tut un lunsdi
> E al demain et tresqu' a mercresdi,
> Qu'el n'alaschat ne hure ne prist fin
> Jusqu'al juesdi devant prime un petit,

Que li Franceis ne finent d'envair
Ne cil d'Arabe ne cessent de ferir. [vv. 1122ff.]

In this description of the battle there is no indication of its actual chronological beginning and end which would tell one how long it lasted, nor is there any mention of the temporal framework within which all this took place; instead, the days of the period in question are successively enumerated, so that one lives through it and experiences its consummation. Successive events are depicted as such by the poet.

Other indications of time in the *chanson de geste* betray the narrow temporal horizon of the author. They have no strictly temporal function, but serve to convey atmosphere. An enumeration of years would communicate no such atmosphere; it would mean nothing to the poet and his readers. One's imagination, however, is stimulated by means of an indication of the season, the day of the week and in particular of the time of day. Such indications must be appreciated in and for themselves, for often they bear no relationship to one another. Does it, for example, facilitate our understanding of temporal relations if we read in *La Mort de Garin* that the abbot Liéteris arrived in Saint-Amant on a Monday,[9] or that Garin destroyed the walls of Morlant on a Monday morning?[10] The following line occurs eleswhere in the same poem:

Il ont mangié un po apres midi. [v. 2239]

On the other hand, there is never an indication in a new episode of the epic of the time that has elapsed in the meanwhile. Even such indeterminate phrases as, for instance, 'Many days [weeks, etc.] had meanwhile passed' are as a rule absent. It is thus impossible to gain even an approximate idea of the time taken by the action in the *chanson de geste*.

With Wace, the sense of temporal order and of historical reality stands out more strongly for the first time. The events narrated in the rhymed chronicle do not occur in an unnamed period of time, but show a definite connection with a particular juncture in history. He was the first to date historical events:

Wit cenz ans e seisante e sis out trespassez,
Puis que Deus de la Virge en Belleem fu nez,
Quant Rou fü a Renier al Lunc Col acordes.
[*Roman de Rou*,[11] II, vv. 393ff.]

The predominance of absolute temporal perspective is linked with the absence of objective time. The poet feels time in the same way as his heroes. He lives with them to such an extent that he lives in *their* time; this enables him to represent, with inward conviction, events that have already taken place as future events. He is able to forget his own position in time—so much so, indeed, that the part of the story still to come takes on the quality of the actual, unknown future of real life. 'The poet becomes involved in the vortex of the manifestations of the desires and emotions of his characters to such an extent that he becomes oblivious of the outcome of the story—which, of course, is nevertheless known to him.'[12]

His awareness of the temporal relations governing the action or events becomes clouded. The *consecutio temporum* in speech is not rigorously observed. The author of the *Chanson de Roland* wavers indiscriminately between present and perfect. His temporal orientation is purely emotional, subjective, and void of any sense of temporal distance. His eye is focused on the 'how' of events rather than on the 'when': 'Il est de règle, du reste, que le mode l'emporte sur le temps.'[13] Circumstances are not viewed externally, objectively and with regard to their duration, outcome and effect, but instead the author becomes subjectively involved in them and experiences them accordingly. There is therefore no awareness of the influences which circumscribe the action and restrain the play of emotion within the action. The poet does not stand at any distance from it and see it as set out in time. He does not view succession spatially and conceptually, but experiences it as it were rhythmically. In the verb forms of Old French the significance of the action as expressed by the verb is stronger than the indication of time implicit in the finite form. Pure tense is not sharply defined, either by verb forms or by the significance of individual words. Temporal adverbs, conjunctions and prepositions are employed sparingly. The periods of the *Chanson de Roland* show an extremely tenuous temporal relationship with one another. The sentence stands in isolation, without any link with the preceding one; nor does it prepare the ground for the next. The poet's thoughts seem to be bounded by the content of the sentence he is uttering at the moment.

Time as transitoriness

The Middle Ages were familiar with the concept of transitoriness inculcated by the Church. There is, however, another interpretation of transitoriness, which is no divine revelation, but scientific and historical understanding derived from experience; everything passes away in nature and in history. From this elementary experience man may most readily gain an awareness of time, by virtue of the latter's manifest effects. But even these effects were not perceived by the time-blind individual, with whose idea of the world we are familiar from national epic poetry. He was aware neither of the falling of leaves in the autumn nor of the passing away of generations. These were phenomena which in no way attracted his attention. The essential quality of the world was its transitoriness *vis-à-vis* God, not the visible change which went on unceasingly in the world.

Man had as yet no eye for the panorama of the centuries, nor had he any comprehensive view of the succession of events. For him, only a particular point of time was illuminated, and what followed or preceded it was lost in darkness. Time was no problem. Its passing was a fact so axiomatic and so essential that one was not even aware of it. It was a part of man himself, and in his imagination it was not a property of things; it lacked both dimensions and proportions. It was experienced, but nothing was known of it.

Nowhere in the *chanson de geste* do we find an observation concerning time. It was not yet a subject for reflection. We look in vain in the national epic for evidence of any attempt to take up a philosophical or moral position with regard to time. The concept of 'time' in the general and comprehensive sense of 'duration' played virtually no part in speech. Time did not exist. In Old French there is no indication of any awareness of an objective time affecting everyone. Everybody experienced his own individualized time, something that passed away with him. In the *Chanson de Roland* 'time' signifies above all *temps à vivre*, and is used almost invariably with the possessive pronoun:

Il est mult vielz, si ad sun tens uset. [v. 523]

Ki ne s'en fuit, de mort n'i ad guarent:
Voillet o nun, tut i laisset sun tens. [v. 1419]

Morz est li quens, de sun tens n'i ad plus.   [v. 1603]

Ço sent Rollant, de sun tens n'i ad plus.   [v. 2366]

In every language 'time' is apt to acquire the meaning of 'life'.
This may readily be understood, for the 'life-time' of a human
being is the time which affects him most strongly, and which
occupies a place in the forefront of his consciousness. It is an
element of one's self so essential that without this 'time' no other
conception of time can exist. It may be said that in early French
thought the concept of time was coterminous with that of life,
for one's interest in the former was purely practical, instinctive
and unspoken. In *La Vie de Saint Alexis* we find the word *tens*
used in the sense of 'epoch' or 'age': 'Al tens ancienur';[14] 'Al
tens Noe'.[15] Here the idea of possession is unclearly expressed.
Time was still considered as individualized; there were as many
times as there were men. Thus we find the plural of *tens* in the
very earliest writings. The stylistic use of the word was still
confined to a very narrow sphere. Phrases such as *je n'ai pas le
temps, tuer le temps, perdre le temps, gagner du temps, profiter du
temps, réparer le temps perdu, avec le temps*, and so on, belong to a
later era. In the *chanson de geste* even the elementary observation *le temps passe* is absent.

The sense of time had to be objectified and the temporal
horizon extended if the idea of transitoriness was to be grasped.
The relatively small sections of time in which the happenings in
the national epic were as a rule set out did not encourage this
sense of transitoriness. A temporal perspective that embraced
centuries and apprehended past things as truly past—over and
done with, no longer existing—is first found in the writings of
Wace, a Norman. He knew what transitoriness was:

Tute rien se turne en declin,
Tut chiet, tut muert, tut vait a fin;
Hom muert, fer use, fust purrist,
Tur funt, mur chiet, rose flaistrist,
Cheval trebuche, drap viellist,
Tute oeuvre faite od mains perist.

[*Roman de Rou*, I, v. 65]

He was the first to recognize changes brought about by time:

Par lunc tens e par lungs aages,
E par muement de langages,

Unt perdu lur premerains nuns,
Viles, citez et regiuns.   [*Ibid.*, I, v. 77]

Thus, time is first apprehended as transitoriness, without (as is evident from the first of these two excerpts) this experience of impermanence being designated as 'time'. In the eyes of the poet it was things that passed away, not time. As we have seen, 'time' signified this or that section of time and not, as yet, that reality embracing all impermanence. A linguistic and intellectual step forward was taken when Marie de France wrote:

Cum plus trespassereit li tens,
Plus sereient sutil de sens.
E plus se savreient guarder
De ceo qu'i ert, a trespasser.[16]

The fact of time's passing was a relatively recent addition to thought. Old French literature was familiar with the process of transition only from units of time: *passet li jorz*. The constatation that a day or a year had passed was incidental and determined by the course of the narrative: the word generally employed to denote the flow of time was *passer* or sometimes *trespasser*. Time passed, to be sure, but its inner melody, its unceasing whispering and flux, was still beyond man's intellectual grasp. In his eyes the world was too rigid and immovable, and perceptions were too inhibited to enable that becoming and passing away which we embody in the image of flowing water to be experienced as time. And yet we find this image as early as the first part of the *Roman de la Rose*, in which Guillaume de Lorris strives to define the nature of time in a passage thirty verses long:

Li tens qui s'en va nuit et jor,
Snas repos prendre et sans séjor,
Et qui de nous se part et emble,
Si céléement, qu'il nous semble
Qu'il s'arreste adès en un point,
Et il ne s'i arreste point,
Ains ne fine de trespasser,
Que nus ne puet néis penser
Quex tens ce est qui est présens;
S'el demandés as clers lisans,
Aincois que l'en éust pensé,
Seroit-il jà trois tens passé;

B

Li tens qui ne puet séjourner,
Ains vait tous jors sans retorner,
Com l'iaue qui s'avale toute,
N'il n'en retourne arrière goute;
Li tens vers qui noient ne dure,
Ne fer, ne chose, tant soit dure,
Car il gaste tout et menjue;
Li tens qui tote chose mue,
Qui tout fait croistre et tout norist,
Et qui toute use et tout porrist;
Li tens qui enviellist nos pères,
Et viellist rois et emperières,
Et qui tous nous enviellira,
Ou mort nous désavancera;
Li tens qui toute a la baillie
Des gens viellir; l'avoit viellie
Si durement, qu'au mien cuidier
El ne se pooit mès aidier,
Ains retornoit jà en enfance.   [vv. 361–91[17]]

Whereas in the quoted verses of Wace the visible effects of time,
and, to a lesser extent, time itself, are described, the first poet of
the *Roman de la Rose* comes closer to the nature of time. He
perceives not only the results, but also the constant becoming of
time—its dynamism, its flow. He is aware of the powerlessness
of man with regard to it, of his inability to stay its current or to
arrest its noiseless flux. It is remarkable that the thought of
death neither gave rise to this view of impermanence nor exerted
any appreciable influence on its nature. The image of corporeal
dissolution and decay was a creation of the later Middle Ages,
with their more acute sense of visual impressions. In the age of
the Crusades transitoriness was still not a contrast but a melody.
Man was still unaffected by the painful awareness that he himself was a victim of this impermanence. He spoke of it as a child
would speak of death. He saw the ship sailing, but he was not on
board it. The passing of time was not yet experienced in the
shape of the brevity of one's own life. It was realized, without
doubt, that the years passed, but it was not yet said that they did
so rapidly.

Human perception was still far from seeing the shortness of
life in that perspective implicit in such phrases as 'Hier j'étais
jeune' or 'Le berceau touche à la tombe'. Before the time of

François Villon the French consciousness was innocent of this mode of thought. The idea of impermanence had nothing tragic about it, either here or in the imagined world beyond the tomb. Impermanence had only the value of all earthly things. It was only a phenomenon of this world. The truly valuable was not what came and went but what is, was and always would be.

Even as the Normans were the first to become aware of time, and even as the Norman Philippe de Thaun was the first to concern himself with the reckoning of time in the popular tongue, so the reproduction in language of a temporal rhythm, a conscious melodic reproduction of temporal relations, unknown either to the *chanson de geste* or to Chrestien de Troyes, was the work of this race. The march of events was rhythmically depicted in speech, evoking the ticking of a clock or the regular falling of drops of water. In the following lines Wace portrays the dissembling of the Norman duke, Hastings:

> Suuent iert pale, suuent ert pers,
> Suuent as denz, suuent envers,
> Suuent s'endort, suuent s'esveille,
> Suuent s'estent, suuent ventraille.   [*Roman de Rou*, I, v. 578]

Marie de France employed the same stylistic device:

> En cel tens tint Hoels la terre,
> Sovent en pais, sovent en guerre.   [Guigemar,[18] v. 27]

In these excerpts one discerns the author's intention of making a temporal observation, implicit in the idea of 'often', not merely conceptually, as a simple statement, but also sensually, in a rhythmic manner. The Norman ear seems to have been particularly receptive to the rhythm of the pendulum swing, to a regular to-and-fro in the melody of circumstances. In like manner Marie de France musically 'sensualized' the idea of 'always':

> Et nuit et jur et tost e tart
> Ele l'a tut a sun plaisir.   [*Yonec*, v. 226]

Wace stylized the alternation of day and night as an acoustic experience; both appear as pendulum swings of equal duration:

> Franceis distrent as noz que as lur n'asemblassent,
> Par els fussent le iur et par els chevalchassent,

Par els fussent la nuit et par els herbergassent,
Que estris ne creust, ne que ne se medlassent.

[*Roman de Rou*, II, v. 1646]

One means, unknown to Old French, of giving rhythmical expression to the idea of time is the repetition of the verb in the same tense, etc, with the purpose of reinforcing the impression of duration one derives from a long-drawn-out experience, as, for instance, in: 'La flauta suena y suena . . . este viejo telar marcha y marcha con su son rítmico' (Azorín, *Una flauta en la noche*). This device seems to have been first used in the era of the Renaissance:

Va, pleureuse, et te souvienne
Du sang et de la playe mienne
Qui coule et coule sans fin.[19]

The subjective and personal conception of time

The absence of contact between the mind and the concept of time, the latter fulfilling the function of an external necessity and human convention, did not show itself merely as a deficiency but in a positive manner in the meaning and use of words, wherein a strong personal and subjective element evidenced itself. The idea of duration, as we understand it, is devoid of any emotional or graphic significance. The fact that anything endures means for us only that a process has a certain temporal extension. Old French imparted sense and feeling to this purely temporal attribute. The words *durée* and *durer*, like *tens*, often had the meaning of 'life'. The general concept of duration was readily applied to an individual—a person 'lasted'—and in this way it acquired emotional overtones:

Ja ne puis durer sens vos:
Et sens moi coment dures vos?[20]

Senz amor ne puis durer ne je ne vuell.[21]

S'amors—ne mi laisse durer.[22]

Ainc n'acointierent li nos pieur jornee,
Car a tes gens avront poi de duree.[23]

The significance of the word passed through the stages: duration–endurance–power of resistance, that is to say, from the temporal to the spiritual. The word was understood as conveying not the general concept but the concrete particularity. In the same way, the idea of 'staying' was foreign to the French spirit of early times. 'To stay' has temporal and spatial connotations denoting an external circumstance: 'The king is staying three days in this town.' 'Why is he staying here?' 'What will he do during this sojourn?' These questions, with their affective overtones, were more in keeping with the mentality of those times, which laid greater stress on inner experience than on the when and where. In Old French the word *sejorner* generally meant 'to rest' (*un destrier sejornet*, 'a rested horse'). The mind was so preoccupied with the import of the sojourn and so little attention was paid to its duration that *sejorner* and *sejor* were very seldom used to indicate duration. In the later centuries of the Middle Ages, when temporal relations were more clearly discerned, the word *sejorner* was increasingly often used to indicate duration, becoming sharply differentiated from the idea of 'rest'. The following example will serve to demonstrate that it had, in Middle French, no longer anything to do with 'repose' or 'tarrying':

> Et de là allay à Millan, où pareillement sejournay deux ou troys jours pour leur demander des gens d'armes.[24]

The same considerations applied to *demorer*, which did not as yet have the meaning of 'to dwell' but rather that of 'to remain'. But this word also was used in a stylistically circumscribed sense, for it was generally used without any indication of place. Typical examples of its use are the following:

> Culchet sei a tere, si priet Damnedeu
> Que li soleilz facet pur lui arester
> La nuit targer e le jur demurer. [*Roland*, vv. 2449–51]

> La noit demurent tresque vint al jur cler. [*Ibid.*, v. 162]

> Ço qu'estre en deit ne l'alez demurant. [*Ibid.*, v. 3519]

Demorer was linked more closely with time than with space. The spatial aspect of 'remaining' and 'tarrying' was unimportant. In the *chanson de geste* there is no clear picture of external circumstances. Given the emotionalism informing the *Chanson de Roland*, we find that, in those passages where there is much movement, 'to remain' usually means 'to remain behind', an action rather than a state, and free from the idea of spatial distance. *Demorer* still had strong affective overtones, denoting spiritual activity. It was not the place where the action was going on that was important, but rather where it was tending and its indeterminate temporal extension. Those adverbs employed to qualify *demorer* are indications of subjectively felt duration.

I out tant demoret.	[*Voyage de Charlemagne*,[25] v. 214]
si ai molt demoret	[*Ibid.*, v. 218]
E cil est loinz, si ad mult demoret.	[*Roland*, v. 2622]

Such indications convey the attitude of the one remaining rather than the act of remaining itself. What has been said of *demorer* applies also to other words denoting the idea of remaining, such as *remaneir* and *ester*. The essential feature of an action was the mental attitude underlying it, rather than its extension in space and time. What was said about time was also said of oneself. The rarity of the phrase *il est tart* demonstrates the extent to which the idea of time as expressed in language was subjective and bound up with the individual. The phrase denotes a temporal circumstance, no matter how subjective this may be, as a fact independent of the judgment of the individual. It is not late for you or for me or for him; it is late without reference to a person. In Old French, however, it was almost always *il m'est tart* or *il me tarde*. This points to a soul-state; something is said concerning the desires, longing and impatience of a particular person. In the case of *durer*, in addition, the personal pronoun is very frequently found in the dative in the language of the eleventh and twelfth centuries, which provides palpable evidence of how little duration was apprehended in general and objective terms:

Tant cum li iurz lui duret.	[*Charlemagne*, v. 245]
Petit m'a dure, Que trop tost sumes desevre.[26]	
Et se çou nos dure longuement, Sire diex, ke devenrons nos?[27]	
Toutes ses armes li ont petit duré.	[*Aliscans*, vv. 5493, 5521]

It would be false to regard this apprehension of time denoted by a personal dative as individualistic. It is true that duration was apprehended in the form of an individual judgment, but this was not a reaction, not self-assertion of the individual *vis-à-vis* objective time, not conscious subjectivism, but a spiritual attitude which had yet to discover the existence of an impersonal time. This impersonal time had so far not acquired the status of an entity independent of personal judgment. If one construes a phrase such as *petit m'a duré* as evidence of marked self-awareness, one must bear in mind that it nevertheless implies a reticence on the part of the ego rather than the reverse. The self lay *under* the objective temporal order, not, as did the awareness of time in the age of the Renaissance, *over* it. This subjective attitude was not a criticism of the course of a world directed by God. This would have been a presumption foreign to the spirit of the Middle Ages.

Impatience

Impatience is a desire, entertained more or less fervently, that a circumstance which has not yet come about should do so as speedily as possible. Impatience, therefore, contains within itself two conceptually distinguishable demands: (*a*) that something should occur; (*b*) that it should do so within the shortest time imaginable. The mutual relationship of these two demands, the absolute and the temporal, may vary, depending on whichever predominates in the mind of the impatient person. Impatience is naturally experienced as an integral feeling; if we are impatient we are not aware of the separate existence of these two demands. We generally measure the degree of our impatience not by the acuteness of our desire to see our wish gratified but by our inability patiently to await its fulfilment. The more ardently we desire to see the object of our wishes, now future, realized in the present, the greater is our impatience. There is impatient expectation that is purposeful and expectation that is, as it were, freely impatient. We have the first when the time of the desired event is seen by our consciousness as unalterably fixed. 'Uncommitted' impatience exists when an element of temporal uncertainty is bound up with the fulfilment of our wish. When the object of our impatience is a fixed point in the future, this point is separated from the present by a definite time span and is thereby linked with it.

This form of impatience is not consistent with the character of the human types we meet with in the ancient French epic. The attitudes we observe are very seldom those of a person who cannot await the passing of an objectively given period of time—which, moreover, he knows he cannot shorten. This attitude was adopted, for example, by Laudine, who in her amorous languishing cannot wait for the day she has appointed as the latest date for Ivain's return:

> Ma dame paint en sa chambre a
> Trestoz les jorz et toz les tans;
> Car qui aimme, est en grant porpans,
> N'onques ne puet prandre buen some,
> Mes tote nuit conte et assome
> Les jorz, qui vienent et qui vont.
> Sez tu, come li amant font?
> Content le tans et la seison.   [*Yvain*,[28] vv. 2754ff.]

This is not the form of impatience we regard as typical of the Frenchmen of those days. The impatient person with a calendar in his hand is a figure unlike the image we form of the individual of the twelfth century. Inability to wait manifests itself more often when there is no anticipation of an appointed date, that is to say, where the event is still temporally 'free'. These are cases in which the impatient person is able to hasten the occurrence of the desired event. It is a highly active impatience, which restricts itself to that which can be attained through one's own action. It is impatience with a sword in its hand, as Shakespeare has it: 'Sheathe thy impatience' (*Merry Wives of Windsor*, II, iii).

The act of desiring and its object are often difficult to distinguish in consciousness; they are frequently experienced as one and the same thing, so that, in Old French, desire and fulfilment were represented as being temporally unseparated ('to wish to fight' = 'to fight'). 'I now desire that something shall happen in the future' was less characteristic a turn of phrase than was 'I shall desire that something shall happen', so that desire and realization went hand in hand. This is probably the psychological explanation of the use of *voloir* in the following examples:

La vuldrat il chrestiens devenir.   [*Roland*, v. 155]

Jo sai mentir, si li voldrai cunter
Que je vus sui par force eschapez.   [*Guillelme*, vv. 1536-7]

Devant l'autel faites aparillier
Un riche lit ou me volrai couchier.
Au crucefis me volrai apuier,
Et les nonnains prendront mi esquier.
                    [*Raoul de Cambrai*,[29] vv. 1238ff.]

The idea of volition was so indissolubly bound up with the desired action that volition was possible only in cases where immediate realization was feasible:

Ferir n'en volt, se n'en fust desturnet.   [*Roland*, v. 440]

The impatience of these people did not spring from an awareness of the need for haste: it was ignorant of the price of time; haste and agitation emanated from an inner constraint. With these people, haste did not arise from an understanding of the state of things or of the exigencies of the hour. It was not until the later Middle Ages that the need for pressing ahead was represented as having its source in externalities. Since any clear vision into the future was still lacking, impatience was all the more urgent and desires clamoured for immediate gratification. There was no such thing as expectancy on a long-term basis. In Old French, *attendre* either referred to a quite short time span or else (and most frequently) it was used in the negative: 'Ne peumes atendre',[30] 'Tres qu'al jur n'atendum',[31] 'Trop n'i as atendu',[32] 'Plus n'i as atendu'.[33]

This inability to wait and the restless desire to act had their own ethos. They were prized as a positive mental attitude.

Or ne quidiés mie que j'attendisse tant que je trovasse coutel dont je me peüsce ferir el cuer et ocírre. Naie voir, tant n'atenderoie je mie.[34]

The unbridled impatience of the epic figure was one of the most striking attributes of the heroic turn of mind. In this emotional atmosphere the significance of *attendre* took on a peculiar colour. Very often, waiting was not a state with objective duration but a gesture of anticipation of something about to happen, the expression of tension that demanded immediate relief. Waiting was an *active* state, revealing a distinctly recognizable mental attitude:

Par tans fust buens li ferëiz,
Se cil les osassent atandre.[35]

This active state was often emphasized by means of a prefix— *contratandre*. In the course of the linguistic changes leading to modern French, *attendre* increasingly implied a state with a definite temporal extension. Its purely temporal significance became ever more prominent. 'Waiting' came to signify the external circumstance of the postponement of an action, without the emotional concomitants of striving or renunciation. We must turn to Middle French if we seek more precise indications of the idea of duration bound up with the state of waiting:

Li diex d'Amors, qui l'arc tendu,
Avoit tote jor atendu
A moi porsivre et espier.[36]

Q'atendu l'ai trois jors enters.[37]

Et atent jusques à cele heure
Qu'il cuida qu'il fussent ensamble.[38]

Or attendez à samedy.[39]

Par aucuns jours attendoit toute la venue de son armée.[40]

Quant ilz eussent attendu deux jours a eulx rendre.[41]

Also linked with the theme of impatience is the shift of meaning of the Old French *souffrir* from 'to suffer' to 'to await', a phenomenon which may further be observed in the Latin *patientia* and the German *gedulden*. A strong desire to be freed from something unbearable is, as we have seen, a temporal demand that this release should come about as soon as possible. The suffering of ill is a temporal acceptance, that is to say, 'to suffer' takes on the meaning of 'to be patient'. This temporal significance is particularly marked in cases where impatience is regarded as a positive moral value, and where temporal acquiescence is considered as an acceptance of circumstances. Those instances in which 'to suffer' is the equivalent of 'to be patient' may readily be divided into two categories: the one consists of cases in which a person's impatience is placated by another, and those which convey a gesture of impatience:

Dist li portier: Sire, un pou vous souffrés.　　　[*Aliscans*, v. 1600]

Dist Rainouars: or vous ales soufrant.　　　[*Ibid.*, v. 5429]

Ja en avrés, or vous alés souffrant.　　　[*Ibid.*, v. 5454]

La dame dist qu'ele est malade;
Del chapelain se prenge guarde.
Sil face tost a li venir,
Kar grant poür a de murir.
La vieille dist: Or suferrez!
Mis sire en est el bois alez.　　　[*Yonec*, vv. 177ff.]

Soufres, maris, et si ne vos anuit:
Demain m'ares et mes amis anuit.[42]

　Suffrez moi
Tant que Adam soit en recoi.　　　[*Le Jeu d'Adam*,[43] v. 273]

Mari vos qier por mon cors deporter;
Or est li termes et venus et passés,
Ne m'en puis mais soufrir ne endurer.
　　　　　[*Raoul de Cambrai*, v. 5789]

After the heathen Aenré has killed fifty Frenchmen, Rainouars says:

Or ai trop enduré.
Se plus i sofre, dont aie je dahé!　　　[*Aliscans*, v. 5828]

Or sui je fox; se plus te vois soufrant.
[i.e. 'If I continue to remain an inactive onlooker.']
　　　　　[*Ibid.*, v. 5958]

Chantecler quide que voir die.
Lors let aler sa melondie
Les oilz cligniez par grant oir.
Lors ne volt plus Renars soffrir.
Par de desoz un roge chol
Le prent Renars parmi le col.　　[*Roman de Renart*, II, vv. 345ff.]

To possess one's soul in patience, therefore, was never represented as a fact but, at most, as something desired of another. We shall see how, in the later centuries of the Middle Ages, this evaluation of impatience underwent a complete *volte-face.* The temporal associations which, as we have seen, were attached to *souffrir* may, in Old French, be observed in other terms expressing keen desire or displeasure. *Ennuyer*, which originally meant 'to vex' or 'to harass', acquired the sense of 'to bore' or 'to be tedious'. *Jaloux* took on a temporal flavour when, as sometimes happened, it meant 'impatient':

> Que que la bele Ydone pleure et plaint et dolouse
> Le preu Garsilion qui tant aime et golouse,
> Atant es vos sa maistre de tost aler jalouse,
> Isnelement courant toute une voie herbouse.[44]

The linguistic resources of Old French, therefore, represented impatience as an inner stress and not as an urge determined by an external agency. Speedy action was motivated solely by the emotional attitude of the doer. A word denoting 'haste' might, on the lips of a medieval poet, acquire a meaning expressing an inner urge, as in the following example:

> Li quens Guillames se hasta de l'entrer.   [*Aliscans*, v. 1645]

Guillaume d'Orange, on his return from the battlefield, was denied admission by his wife. The sentence does not mean 'he made haste to enter' (indeed he could not, for he was refused admission) but 'he felt an urge to do so, he keenly desired to enter . . .'

Emotionalism, time and space

It may appear contradictory if we observe, on the one hand, that in Old French linguistic usage the emotional aspect is stressed in temporal concepts, and, on the other, that the temporal element is emphasized in words such as *souffrir* and *jaloux*. These seem to be two diametrically opposed linguistic, and hence emotional, tendencies; but this is not so, as we shall demonstrate.

From the standpoint of our inquiry there are three possibilities as regards meaning in words: (*a*) a purely emotional one—'I wish'; (*b*) a temporal and emotional one—'I feel an urge to do something'; (*c*) a purely temporal one—'I hasten'. For (*c*), (*b*) and (*a*) we might substitute respectively: 'it is late'; 'it is late for me', that is to say, I feel pressed to do something; or 'I wish', with no temporal implication in the meaning. In the examples we have taken from Old French, we notice a tendency to move away from the purely temporal on the one hand, and the purely emotional on the other, towards an intermediate blend of the two, the emotional–temporal. The linguistic facts we have observed show that meaning might move in various directions, but always towards a common terminus, beyond which changes of meaning in Old French did not pass. *Souffrir* changed its meaning from 'to suffer' or 'to bear' to 'to be patient', but never to the purely temporal 'to wait', without psychological overtones. Purely temporal concepts, the purpose of which is simply to establish relations in time, were alien to the turn of mind of those days. The popular tongue (and indeed not only the *chanson de geste*) strove to inject feeling into purely temporal indications, that is to say, it increasingly aimed at conveying individual emotion and volition rather than time pure and simple. Leo Spitzer made a similar observation: 'The purely temporal future was foreign to the popular spirit.'[45]

The Romance future certainly did not owe its existence to the need for a new tense. Only with the progressive emancipation of thought from popular habits did temporal indications and the conveying of emotion become differentiated in the significance of words. By virtue of the necessity of order created by civilization, time became ever more important and autonomous as a determining element, and increasingly became an independent entity divorced from life. We may observe how, in the languages of civilized countries, the meaning of words which originally had both a temporal and an emotional connotation shed this combined association and became either purely temporal or purely emotional. The Germanic word *bald* ('soon') acquired in German a purely temporal, and in English a purely psychological, connotation. Old French contained many expressions in which the significance of words denoting speed was still bound up with the denotation of personal qualities. Indications of speed did not as yet refer to the action itself, but to the worth of an individual. Tardiness, for instance, appeared in some sort as a property of the soul:

Margos le lieve, ki n'ot pas le cuer lent.   [*Aliscans*, v. 5744]

Space underwent the same development as time in the modification of language and thought. Space, no less than time, was originally not isolated from the experience as a whole, but ultimately it was ever more sharply differentiated from time and from the emotional content of the experience. 'Even as space, inseparable as it was from the concrete fulness of events, was for this very reason closely bound up with time, so, conversely, all temporal elements could be grasped only in the concrete totality of the occurrence, and consequently they were combined with spatial elements as an undifferentiated homogeneous entity.'[46] In Old French, moreover, the spatial and the temporal were not as yet firmly and sharply demarcated. This is not to be understood as implying that temporal concepts were projected into space. Time and space alike were absent from mental images, for it is impossible to become aware of either unless they are distinguished, the one being emphasized to the total exclusion of the other. The word *saeculum* (O.F. *siecle*) was a case in point, for not until later did it acquire the distinct meanings of 'century' and 'world', temporal in the one case and spatial in the other.

The progressive separation of the two spheres was resisted by the poetic synthesis, which was again and again interposed between the two by the popular tongue and the spirit of the poet. In the pages of Shakespeare 'time' has not the meaning of pure succession; it is not on the plane of the temporal order but combines time and space in an indissoluble unity. 'Time' is for the poet the world: external happenings and inner experience—the entire content of consciousness. With him, time becomes the mundane order of things. He uses the word 'time' to denote the world, in its state of change and flux, as often as the word 'world' itself, because for him they are the same thing. In his view the world is a process, and duration, as it were, a theatre. In the Tower, Clarence speaks of his dream thus:

> I would not spend another such a night,
> Though 'twere to buy a world of happy days;
> So full of terror was the time.　　　　　[*Richard III*, I, iv]

In contradistinction to this, civilization, with as it were a sort of centrifugal force, drives apart the significance of concepts in opposing directions, those of time and space. It may be that the differentiation between *derrière* and *après* owes its existence to this tendency. The antithesis between 'in front' and 'behind' is no purely spatial one. Of the three principal dimensions (in front–behind, above–below, left–right), it is the first-named that has the strongest temporal content.[47] In Old French the demarcation between the temporal and the spatial meanings of these words had not been fully established. Originally, 'behind' and 'after' were one and the same, the former being associated with space and the latter with movement. Events imagined as happening later seemed to the consciousness to take place elsewhere. In French the words *avant* and *devant* became differentiated in their functions in the course of their development. In the medieval tongue this possibility of a difference in meaning was not realized. In English, likewise, the words 'nearest' and 'next' have become separated; one says 'the nearest way' as against 'Monday next'.

It is not the case that, of these pairs of concepts with closely associated meanings, one word has a strictly spatial and the other an exclusively temporal significance. It would be more accurate to speak of the degree and gradation of their inherent temporal or spatial content. *Suivre* and *succéder* are cases in point. The former is the more comprehensive in meaning, because it is the older, whilst the latter remains confined to its temporal sense. *Suivre* may have now a temporal, now a spatial connotation; *s'ensuivre* is on the temporal plane. *Avancer* and *devancer* show, compared to *précéder*, a more dynamic, more spatio–temporal shading. The co-existence of such words allows a variation of expression according to its temporal content, a variation unknown to the language of the *Chanson de Roland*.

The future

The attitude adopted by a culture towards the future is of importance for the understanding of its true nature, for it may be that an epoch's general feeling for time is most clearly reflected in its consciousness of the future. Thus Emerich Madzsar considered the essential feature of the Babylonians' concept of time to be the special enthusiasm with which they turned their attention to coming things and endeavoured to establish a foundation for them on the basis of their belief 'that every event or thing significantly pointed to the future, or, conversely, that everything that would happen in the future was predetermined and implicit in present auguries'.[48] This 'predictability'[49] was not a characteristic of the consciousness of the future peculiar to the Middle Ages. For an individual possessed of an essentially objective conception of time, at all events, there can be only a slight modal differentiation between what is present and what is to come. For him, the future falls into days, months and years as much as do the past and the present. All times, for him, are sections of a straight line divided into fixed units. He 'thinks' the future into units of time, treating the past, moreover, in the same way. To a person with this turn of mind, time lying both before and behind him appears essentially the same. The image of the future which his view of things may project is not the future at all, but the present; accordingly, the saintly Augustine designated that time which is only imagined and which has not become reality as *praesens de futuris*;[50] and the philosopher, who believed himself exempt from all emotional impulses, said: 'It is a matter of indifference whether that time which does not fill our existence should acquire, relatively to that which does, the form either of the future or of the past'.[51]

But the individual who evaluates time as an experience does not entertain this indifference and eschews such levelling-up. Everyone is preoccupied with the question of survival after death, whereas pre-existence is a matter for the philosophers. This difference in the individual's relationship to the past and the future may also be formulated genetically: 'The relationship of the consciousness to the future develops earlier than its relationship to the past.'[52] This phrase, inspired by the development of the individual consciousness, applies also to the history of a culture. The past and the future are differently evaluated and sensed. The person who experiences time turns to the past in a different frame of mind to that in which he considers what is to come. For him, they are not only different times, but different aspects of his being. The future is not simply a continuation of what has been and what is, as it seems to be to our spatialized conception of time. It is true that the complete divorce of the present from the future is an extreme contingency which never arises in the reality of human emotional life. With every person, numerous and intimate links exist between the two.

Here mention may be made of our ability to become aware of 'time shapes', that is, to apprehend greater or smaller periods of time as instants. A familiar tune is one such 'time shape'. 'If, for instance, I have grown to know a tune, and if I hear someone repeatedly intone it, the entire melody is in some sort contained within the initial notes. The notes actually sounding "represent" (i.e. make present) the whole composition, including those portions of it that are yet to be played or sung.'[53]

A person's relationship to the future is thus essentially determined by his experiences. In the medieval consciousness these were far from performing the function which they carry out in modern thought. One of the most outstanding features of the medieval turn of mind was that it left out of account, to a degree almost incomprehensible to us, the concept of experience, which lies at the heart of the modern understanding of reality. For the mentality of the Middle Ages, therefore, the future was a world of unlimited possibilities; in the circumscribed and certain conditions of the present there was no means of foreseeing what might befall. The sense of the future was strongly inclined towards the state which Hans Keller called *inhaltlich amorph* ('amorphous in content').[54]

The course of the world was unpredictable; it was not as yet controlled by laws to which future happenings were also subject. Events could be known only in so far as they had taken place. They went to form experience, which enables us to judge future probabilities, only to an inconsiderable extent. The future was thus informed with its essential irrationality. The world was full of wonders, surprises, adventures and secrets. This being so, life was much less well ordered, and the future much more uncertain, than they are today. With such an outlook, humanity was apt to believe that the end of the world might be imminent. In a world where natural processes were regulated by no laws, but where everything that happened was regarded as the effect of godly or super-natural intervention, the future was so little conditioned by experience, that people believed that all was subject to divine judgment.

Thus the future might hold anything in store for the humanity of those days, but this 'anything' was, considered from our viewpoint, as circumscribed as were medieval thought and belief. Anything might happen at any moment, but before it could occur it had to pass through the very selective sieve of the medieval interpretative mentality. As an apt comparison, one might adduce the adventures of Don Quixote. Manifold and varied as they were, they all became realities; but one must bear in mind his restricted outlook and the circumscribed range of knightly activities.

The more uncertain the external future appeared, the greater the inability of mankind to obtain some degree of certainty as regards the future, the more absolutely and categorically did it oppose this uncertainty with its own will, the only undoubted factor in this world of unlimited possibilities. The future could not be foreseen, but it was always possible to bring one's own will to bear. In the *chanson de geste* the future had always the connotation of something willed. Will and intention were, with certain reservations, the only things which could lend stability to the future and lessen its uncertainty somewhat. The will was the principal means of establishing a link between the present and the future; if one willed and acted now, one would continue to do so in the future. The element of continuity between today and tomorrow was a matter of one's mental attitude, not a conscious extension of what has been and what is.

Thus we see that those forms of expression having to do with the future pointed less to an indication of time than to a manifestation of the will. This applies above all to the stylistic use of the future in Old French. The Lerchian *Heische-futurum*[55] ('future of desire'), like the *ethicum–advocaticum–oratorium* of Vossler,[56] represented future happenings not only as an image but also as a wish. 'The narrator,' says Vossler of the sudden transition from the perfect to the future tense frequent in Old French, 'assumed a rhetorical, even active, function, not a purely spectatorial one.'[56] The narrator's greatest desire was to gain control of the future. In those cases where the future tense did not indicate active desire it pointed to his lack of insight into time. In the *Chanson de Roland* the future tense tells one more about the present than about things to come. This fact bears out our observation that in Old French, the future tense seldom fulfilled a purely temporal function. The phrase *ja mais*, which in Old French referred solely to the future, was a vehicle of infinite negation as much as of temporal determination. It was the temporally couched form of vast overstatement, serving to denote a present outburst of emotion:

Ja mais n'iert an altretel ne vos face!   [Roland, v. 653]

It is a prerequisite for the understanding of anything expressed in language not to mistake what is said for what is meant. In Old French, one must not allow temporal formulations to mislead one into thinking that the poet was necessarily concerned with time. It was always intensity of feeling that he intended to convey.

The fact that there were, in the ancient tongue, two words expressing the idea of 'never'—*jamais* and *onques*, both used with *ne*, the one employed in respect of future, the other in respect of past events—is incomprehensible to any philological science which thinks and explains in a positivistic manner. Why did a new word come into existence side by side with the old *unquam* and assume the latter's function, at first in part and later *in toto*? The gesture of exaggeration was common to them both, but the emotional situations in which they were employed were different. The element of intensity inherent in *unquam* could grow so strong that it overshadowed the word's temporal significance. Marie de France wrote in the following terms of Equitan, who sprang into boiling water:

Dedenz la cuve salt joinz piez,
E il fu nu e despuilliez;
Unques guarde ne s'en dona.   [*Equitan*, v. 301]

The following example is typical of the normal use of the word:

Tenez mun helme, unches meillor ne vi.   [*Roland*, v. 629]

This is an appeal to the fantasy and power of imagination of the listener; it is an expression of naive, unhistorical admiration of what never existed. The use of the word disappeared with the decline of the medieval belief in miracles. In later and more critical centuries the use of the word *onques* became subject to greater circumspection. Chastellain provides us with an admirable criticism of this idiosyncrasy of the ancient tongue. In his *Temple de Bocace* the Queen of England (Margaret of Anjou) gives vent to her grief to the author of *De claris mulieribus*, making exuberant use of the word *onques* as she does so.

> Comment Tristeur? Est-il en terre qui en ait plus que moy, et qui oncques l'eust samblable à moy, ne à qui tant fust cause de la monstrer par parole?

Thereupon Chastellain puts the following matter-of-fact explanation into the mouth of the Italian:

> Le 'oncques' présuppose un long temps à faire question sur les infimes personnes: entredeux il y en a eu des tristes.[57]

The word *jamais* contains, rather, a volitional element, which can express itself in a negative way also, denoting pain and regret at the non-fulfilment of what is willed. *Onques* excites astonishment, having an epic quality, *jamais* has a dramatic effect, coming more frequently from the lips of the hero than those of the poet. *Onques* expresses personal disinterestedness, the eye being more concerned than the mind. *Jamais*, however, connotes subjective involvement, sweeping one along in its reference to coming things. The contemplative *onques* could not refer to the future, for this is only an object of the will. If it was desired, in Old French, to express the concept of 'never' in its general application to both the past and the future, then *onques* was used in conjunction with *ja*:

> Unques ne fu ne ja nen iert
> Ne n'avendra cele aventure,
> Qu'a une sule porteüre
> Une femme dous enfanz ait.   [*Le Fraisne*, v. 38]

The will, as it projected itself into the future, was at that time essentially impulsive and planless, with no clear picture of the goal to be attained. It was a will uninformed by experience, incapable of estimating the probable course of future events. The calculating 'prudence = providence' did not inform the attitude of mankind towards time until later. Were not the Crusades enterprises undertaken with a great effort of will but little foresight?

One may best understand the co-existence of *onques* and *jamais* if one regards them, not as two expressions of the same idea, but as words denoting two quite different things, two different facets of the soul, two essentially dissimilar gestures orientated in opposing directions. Our concept of 'never', with its stark absoluteness and total encompassing of all temporal possibilities, is an obstacle to our understanding of Old French usage, which reflected the difference between future and past experience more clearly and with greater originality than does our word. This difference of attitude towards things bygone and things to come was not always in evidence when philological distinctions were drawn between the past and the future. When Malherbe, in the seventeenth century, decided that *autrefois* should refer only to the past, and that *un jour* should be employed only with regard to the future, it was a case quite unlike that of *onques* and *jamais*. The philological purist has in view the logical clarity and the precise range of meaning of words. Two words should not fulfil exactly the same function. Malherbe, in determining the use of these words, made a conceptual distinction between past and future. He ordained that, on the straight line representing time, the word *autrefois* be used up to the point of the present, and that *un iour* be employed from there onwards.

The day as lived

The notion of 'the day' has its place in the concept of time considered historically. Hermann Fränkel employed this notion as a means of throwing light on the peculiar characteristics of, and changes in, the ancient Greek conception of time.[58] The day was above all a phenomenon characterized by light; it was the antithesis of night; it was a time which seemed to pass away quickly or slowly, and which always made a strong impression on the unsophisticated and poetic temperament. It came and departed, and one became particularly conscious of it as it dawned and died away. Verbs of activity and movement served to express this experience of the day: to 'move on', 'elapse', 'pass away', 'hasten on', 'slip away', 'draw to a close', 'die', 'dwindle', 'flee', 'flow' and so on. The visual aspect of the impression was conveyed by adjectives of brightness and colour: 'bright', 'clear', 'radiant', 'fair', 'beautiful', 'blue', 'golden' and the like. The poetic mentality did not apprehend the day as an element in a fixed temporal relationship. It took on a personal and emotional hue and was freely and arbitrarily linked with other concepts, such as brilliance, light, sun, beauty, life, youth, impermanence and so forth.

But the day of light could pale into an arithmetical day of twenty-four hours and be no more than a unit of duration. If this happened, immediate concrete experience of time acquired the form of an objective, spatial view of time. The day was no longer an experience and an event, but a unit on a scale. As a precise element of duration it stood in a fixed relationship with minutes and seconds, on the one hand, and to weeks, months and years on the other. It was now an important adjunct of civilization; it was no longer a free, unrelated experience, but a measure, a means of furthering the human need of order—indeed, an organizational institution. In this sense it had no value of its own. It was necessary, but it was not enjoyed, experienced or celebrated. In lyric poetry the day was master; in the ordered intercourse of humanity it was a servant.

The original conception of the day was that of the day of light; its equivalents in other languages—the Greek *νυχθήμερον*, for example—bear this out. 'Day and night . . . are short units which immediately become obvious. Their fusion into one single unit, the day of twenty-four hours, did not take place till later, for this unit as we employ it is abstract and numerical; the primitive intellect proceeds upon immediate perceptions and regards day and night separately.'[59]

The manner in which 'the day' was apprehended in the age of French chivalry is a question we shall answer by examining the philological and stylistic treatment which this word underwent in Old French. In doing this we are confronted by another problem: how did it come about that the Latin *dies* was replaced by *diurnum*? Did a difference of meaning exist between these two words at the time when the latter already signified 'day', and which ultimately caused it to prevail? It might be thought that this word ousted *dies* by virtue of the fact that it designated the perceptible and experienceable day more pronouncedly than did *dies*. It is conceivable that *dies*, which in later Latin and in the language of the Church had acquired associations bound up with authority, civilizing influences, the reckoning of time, and which took on an often specifically ecclesiastical nuance, was ultimately felt by the broad mass of speakers to be an inappropriate term for the day which they regularly lived through. It may be that, in the earliest stages of the French language, there existed an awareness of the stylistic difference between an abstract and official and ecclesiastical day: *dies paschalis, dies dominica, dies sabbati, dies naturalis, dies passionis*, etc, on the one hand, and the natural day of light on the other. By reason of such predominantly date-determining and referential usage, *dies* might well have lost much of its association with 'daylight'.

Certain facts seem to bear out our assumption: the Latin *diurnus* is, by reason of its form, the counterpart of *nocturnus* and therefore more strongly suggested the idea of daylight and the day as actually lived. In late Latin *diurnus*, it may be noted, carried the meaning of 'period of daylight':

Jubet etiam dari vinum noctibus, diurnis, atque jugiter.[60]

In those Old French texts where *di* and *jor* occur side by side, the latter conveys the idea of 'the day' more pictorially than does the former. In the most ancient literary and linguistic monuments[61] of Italy a similar dichotomy may be discerned: that of *dì* and *giorno*, the former being employed in documents whereas the latter was preferred in poetic language. The official, date-assigning, matter-of-fact and commercial word was *dì*. Remarkably enough, it is used preponderantly in the plural in the quotations recorded by Monaci—'tredici dì anzi kalende luglio' (*Libro di Banchieri Fiorentini*, 1211), 'ed ebelo tredici die a l'escita d'otobre' (*Ricordi di Matasala senese*), 'oto dì o un meso' (*Lo splanamento di Patecchio*), to quote but a few examples. On the other hand, *giorno* occurs but seldom in the plural in these texts. Its use probably served to denote the individual day, its antithesis to night-time and its subjective duration. The first recorded instance of its use is in the 'Ode to the sun' of Saint Francis, in which it occurs as a verb:

Spetialmente messor lo frate sole,
Lo quale jorna, et allumini per lui.

The following are typical examples of its use:

Quando lo giorno appare scuto li dolci amori. [Rinaldo d'Auino]

S'io dormo, in mia parvenza,
Tutora l'agio im ballia,
E lo giorno m'intenza,
Di llei sembianti m'invia.   [Canzone anonima, VI]

In the *Divine Comedy* there is no longer any trace of this distinction.

As early as the age of classical Latin, *lux*, with its numerous derivatives, was employed to a greater extent than *dies* to denote the period of daylight, so that *diurnum* took the place of *lux* rather than of *dies*. It may be said that *jour*, which in Old French did duty for *lux*, superseded *di*, for *lux* is absent from even the oldest French texts. The assumption that *diurnum* owed its victory over *di* to its tangible meaning of 'daylight' is perhaps corroborated by the actual use of these words in Old French. The use of *diurnus* in late Latin provides no indication of its subsequent fate, for this word, when used adjectivally during that period, often took on the meaning of 'everlasting, eternal, permanent'.[62] The most ancient French texts, up to the *Chanson de Leodegar*, continued to make a preponderant use of *di*. In the *Passio Christi* it occurs thirteen times, whereas *jorn* is found only six times. In *La Vie de Saint Alexis* the proportion was reversed: *jurn* seven times, *di* only once (l. 140). In *La Chanson de Leodegar* we find 'unches puis cel di', in the *Passio Christi* 'unc puis cel di' (the A manuscript already had 'puis icel jur'). In *La Chanson de Guillaume, La Chanson de Roland, Le Voyage de Charlemagne*, and in everything that followed, only minimal traces of *dies* remained. It persisted only in assonances and in certain set phrases:

La bataille dur[r] a treis dis   [*Gormont et Isembart*, v. 430]

Puis vesquirent il mains dis.   [*Auc. et Nic.*, 41, v. 20]

Greater longevity was enjoyed by *toudis, tandisque, jadis, puiscedi que* (*Cleomades*) and the names of the days of the week.

How is the decline of *di*, at first employed with relative frequency, to be explained?[63] If we consider the meaning predominantly attaching to it in the most ancient texts, we cannot fail to notice how colourless, general, dry and inexpressive the word had become. Its connotation of 'daylight' had become much attenuated. *Di* was a purely temporal concept, devoid of content.

It served chiefly to fix dates and stretches of time:

Passio Christi
Quatre dis	[l. 31]
Et a cel di que dizen pasches.	[l. 89]
Li fel Herodes en cel di.	[l. 218]
Et al terz di vius pareistra.	[l. 364]
En eps cel di.	[l. 417]
Quaranta dis.	[l. 449]

In many cases the word had taken on an even more general and colourless signification, meaning nothing more than *tempus*:

Chi rex eret a cels dis soure pagiens.	[*Eulalia*, v. 12]
Si cum prophetes anz mulz dis canted aveien	[*Passio Christi*, v. 27]
Anz petiz dis.	[*Ibid.*, v. 29]
Qui donc regnevet a ciel di.	[*Leodegar*, v. 15]
Cio fud lonx dis que non cadit.	[*Ibid.*, v. 231]

This usage no longer conveyed the immediate experience of 'the day'. *Di* served for the measurement of some extraneous element, but it was never experienced as a state or an event in itself.

This colourless concept stood in sharp contrast to the increasingly frequent use of *jour* in the *Passio Christi*; the word had already superseded *di* in the *chanson de geste*. *Jour*, no less than *di*, was used for the purpose of fixing the date, but the centre of gravity of its significance did not reside in this use. It was, rather, the word employed to communicate experience of the day. It is evident that a new spirit of love of the day was bound up with the fate of *jour*, a spirit which did not evaluate the day by the yardstick of its objective length or of its practical utility, but which interpreted it *qua* impression and experience. The day was not something taken for granted and allowed to slip by unnoticed. How important was the part it played in ancient French epics, and how numerous were the allusions to its clarity and beauty!

Guillelme:
Clers fut li jurz, et bels fut li matins.	[vv. 234, 1732]
Li soleilz raiet si'st li jurz esclariz.	[v. 235]
Jusqu'al demain que jurz apparut clers.	[vv. 1090, 1565]

Charlemagne:
Li iours fu beaus et clers.   [v. 109]

Roland:
La noit demurent tresque vint al jur cler.   [v. 162]
Par main en l'albe, si cum li jurz esclairet.   [v. 667]
Clers fut li jurz, et bels fut li soleilz.   [v. 1002]
N'unt guarnement que tut ne reflambeit.
Esclargiz est li vespres et li jurz.   [v. 1807]
Carles se dort tresqu'al demain, al cler jur.   [v. 2569]
Clers est li jurz et li soleilz luisant.   [v. 2646]
Clers fut li jurz et li soleilz luisant.   [v. 3345]
Passet la noit, si apert li cler jor.   [v. 3675]

Renaus de Montauban:[64]
Au matin, parson l'aube, quant li jors parut cler. [XLVIII, v. 18]
Ce fu a lendemain que jors fu esclarcis.   [LI, v. 4]
Tantost com l'empereres vit le jor claroier.   [LIII, v. 23]

La Mort Aymeri de Narbonne:[65]
La nuit s'en vet et li jorz esclarci,
Li soleus lieve par estranjes pais.   [v. 2243]

La Mort de Garin le Lorrain:[66]
Li jors fu biax, li solaus esclarcist.   [v. 4113]

Aliscans:
Biaus fu li jors et li solaus levés.   [vv. 3528, 4956]
Li jors fu biaus si prist a esclairier.   [v. 4479]
Biaus fu li jors, li solaus raia cler.   [v. 4601a]
Biaus fu li jors et li tans esclaira.   [v. 4943]
Biaus fu li jors et solaus luist cler.   [v. 5622]

Yvain:
Tant que li jorz fu clers et granz.   [v. 5871]

Guingamor:
Li tens fu clers et li jurs beals.   [v. 333]

Aucassin et Nicolette:
Et se j'atent le jor cler.   [XVII, 1, 11]

These quotations demonstrate that the day was a manifestation
of luminosity that was constantly referred to. All the terms
associated with *jour* served to convey the brightness of the daytime: *clers, bels, matins, soleilz, raier, albe, aparoir, esclairer,
esclargir, luisant.* It may be assumed that there existed a predilection, unknown to later Latin culture, for the natural beauty

of this manifestation. One was as far as possible removed from
the insensitiveness to the beauties of the daytime of the upstart
and debauchee Trimalchio, who expressed his attitude to this
phenomenon in these terms:

> ...dies nihil est. Dum versas te, nox fit. Itaque nihil est melius,
> quam de cubiculo recta in triclinium ire.[67]

In the ancient French epics the day was a splendid experience,
something which imparted its brilliance to every circumstance,
affected everyone's feelings and frame of mind, and played its
part in everything. It illumined the heroic deeds of Roland, it
was mirrored in the gleaming weapons during the battle, and
apportioned to the latter its proper temporal limits. It was the
most impressive phenomenon of nature, seen as it was by the
eyes of the age, concealed and distorted by no veil and by no
clouding of the gaze. But the day was more than an ocular
impression; it was a length of time that was experienced; it was
something that came, passed away and returned. There was no
less awareness of its duration, particularly in the form of an
impression and an experience, than of its radiant brilliance.
When the sun stood high in the heavens the day seemed to be an
eternal state. But in the evening the passing of the day was
noticed and commented on:

Alexis:
Quant li iurz passet et il fut anuited.   [l. 51]

Roland:
Tresvait le jur, la noit est aserie.   [l. 717]
Passet le jurz, si turnet à la vesprée.   [l. 3560]
Passet li jurz, la noit est aserie.   [ll. 3658, 3991]
Passet la noit, si apert le cler jor.   [l. 3675]

In addition, the syntactical use of *jour* was highly characteristic of this interpretation of the idea of 'the day'. Whereas *dies*,
in the examples quoted above, seldom occurs as the subject of a
sentence, because it served chiefly as an indication of time, it was
almost always employed subjectivally in the *Chanson de Roland*.
There was thus a tendency for *di* to be used as an adverb and
jour as a subject. What was said of 'the day' in the *chanson de
geste* did not serve any merely subsidiary purpose; it was itself
the principal object of comment. For in the *Chanson de Roland*

the course of the day was an element which imparted rhythm
and coherence to events: 'the day' was invoked not in order to
establish the dates and times of circumstances but with the
object of stressing that 'the day' was an experience in its own
right. The day was regarded not as a unit of time but as a unit of
experience. It was not only something that might be turned to
account, it was also a fertile source of impressions. The word *jor*
was as yet void of any association with more abstract qualities.
The adjectives used in connection with it in Old French, and
which served to convey its brightness, were *clers, beaus, lons,
granz*[68] and also *antiers*.[69] The idea of 'goodness', for instance,
was not at that time a conceivable property of 'the day'. It was
thus scarcely used as an attribute of *jor* until well into the
thirteenth century. Other adjectives representing an abstract
quality may readily be recognized as Latinisms:

haut:
On pouvoit bien canter et lire
De le sequenche du haut jour.
> [*Li dis dou vrai aniel*,[70] vv. 402–3]

honoré:
Ce fu a Pentecoste, a un jor honoré.
> [*Renaud de Montauban*, I, v. 15]

grant:
Le jor del tres grant vendredi.
> [*Du Chevalier au Barisel*,[71] v. 695]

saint:
Au saint jor del noal.
> [*Aliscans*, v. 548]

festival:
As jurs festivals.
> [*Quatre Livres des Rois*,[72] IV, XVI, l. 18]

These examples demonstrate how the idea of 'the day' was
gradually robbed of its naturalness by the influence of ecclesiastical thought. The religious spiritualization of natural things,
which we observe to take place on all sides in the Middle Ages,
lifted 'the day' on to a higher plane, invested the days with
differences of significance and value, and made of them images
and witnesses of an imposing symbolism.

Notes

1 P. Imbs, *op. cit.*, p. 12. The gulf between Old French and a primitive conception of time may be appreciated by comparing it with an Indian dialect. The Hopi speak a language devoid of temporal abstractions. They are incapable of combining successive time units ('days') with the idea of a sequence ('ten days').—Benjamin Lee Whorf, *Language, Thought and Reality*.
2 Verses 524, 539, 552.
3 Léon Gautier, *La Chanson de Roland*, twentieth edition (Tours, 1892), p. 54n.
4 *La Chançun de Guillelme*, v. 1336 (ed. Suchier, Halle, 1911).
5 Fränkel, *op. cit.*, p. 99.
6 *Ibid.*, p. 102.
7 *Ibid.*, p. 97.
8 Friedrich Schürr, *Das altfranzösische Epos* (Munich, 1926), p. 134.
9 Verse 15.
10 Verse 2593.
11 Ed. Andresen, Heilbronn, 1879.
12 Lerch, *op. cit.*, p. 108.
13 Brunot, *Histoire de la Langue française* (Paris, 1924), I, p. 245.
14 Verse 1.
15 Verse 2.
16 *Die Lais der Marie de France* (ed. Warnke, Halle, 1925), Prolog, v. 19.
17 *Le Roman de la Rose*, (ed. Michel, Paris, 1864).
18 Ed. Warnke.
19 Remy Belleau, *Oeuvres complètes* (ed. Gouverneur, Paris, 1848), III, p. 148.
20 Bartsch, *Altfranzösische Romanzen und Pastourellen* (Leipzig, 1870), p. 36.
21 *Op. cit.*, p. 40.
22 *Op. cit.*, p. 144.
23 *Aliscans* (ed. Wienbeck, Hartnacke and Rasch), p. 263.
24 Philippe de Commynes, *Mémoires*, VI, 4 (ed. Calmette, Paris, 1925), II, p. 271.
25 Ed. Koschwitz–Thurau, Leipzig, 1913.
26 Bartsch, *op. cit.*, p. 33.
27 *Ibid.*, p. 36.
28 Chrestien de Troyes, *Yvain* (ed. Foerster, Halle, 1913).
29 Ed. Meyer–Longnon, Paris, 1882.
30 *Roman de Rou*, II, v. 913.
31 *Ibid.*, II, v. 982.
32 *Aliscans*, CXXI, 28.
33 *Ibid.*, CIII, 46.

34　*Aucassin et Nicolette*, 14, vv. 7–9 (ed. Suchier, Paderborn, 1909).

35　Chrestien, *Cligés*, vv. 1332–3 (ed. Foerster).

36　*Roman de la Rose*, v. 1698.

37　*Roman de Renart* I, v. 443.

38　*Recueil Général et complet des Fabliaux des XIIIe et XIVe siècles* (ed. Montaiglon–Raynaud, Paris, 1872–90), I, p. 195.

39　*Pathelin*, v. 248 (Bibl. Romanica).

40　Chastellain, *Oeuvres* (ed. Lettenhove), I, p. 107.

41　Commynes, *Mém.*, II, 3 (Calmette, I, p. 115).

42　Bartsch, *op. cit.*, p. 20.

43　Ed. Grass, Halle, 1907.

44　Bartsch, *op. cit.*, p. 60.

45　*Aufsätze zur. rom. Synt. u. Stil*, 'Über das Futurum *cantare habeo*'.

46　Heinz Werner, 'Raum und Zeit in den Urformen der Künste', Noack, *op. cit.*, pp. 72–3.

47　William Stern, discussion of W. Morgenthaler, 'Der Abbau der Raumvorstellung bei Geisteskranken', Noack, pp. 95–6.

48　E. Madzsar, 'Raum und Zeit in der babylonischen Kultur', *Archiv für Kulturgeschichte*, 22, p. 75.

49　*Op. cit.*, p. 81.

50　*Confessiones*, XI, 20.

51　A. Schopenhauer, *Die Welt als Wille und Vorstellung*, IV, 41 (ed. Grisebach, II, p. 548).

52　William Stern, 'Personalistik der Erinnerung', *Zeitschr. für Psychologie*, 118, p. 373. Cf. R. Glasser, 'Die Zeit in der Volkssprache', *Indogerman. Forschungen*, 57, p. 187.

53　Hans Keller, 'Psychologie des Zukunftsbewusstseins', *Zeitschr. f. Psych.*, 124, p. 217.

54　*Op. cit.*, p. 221.

55　Eugen Lerch, *Die Verwendung des romanischen Futurums als Ausdruck eines sittlichen Sollens* (Leipzig, 1919).

56　Karl Vossler, *op. cit.*, p. 74.

57　Chastellain, *Oeuvres*, VII, p. 107.

58　Noack, pp. 101f.

59　M. P. Nilsson, *Primitive Time-reckoning* (Lund, 1920), p. 11.

60　Caelius Aurelianus, *De morbis acutis* (Amsterdam, 1755), II, 39, 228.

61　E. Monaci, *Crestomazia italiana dei primi secoli* (ed. Arese, Rome and Naples, 1955).

62　Rönsch, *Semasiologische Beiträge zum lat. Wörterbuch* (Leipzig, 1887–89), II, p. 9; Ahlquist, *Studien zur spät-lat. Mulomedicina* (Uppsala, 1909), p. 131.

63　The problem of the expansion of *jorn* has been treated by Karin Ringenson: '*Dies* et *diurnum*: etude de lexicologie et de stylistique',

Studia Neophil. X (1937), pp. 1–53. Reviewed by G. Rohlfs, *Archiv Stud. Neuer. Spr.* 173 (1938), p. 276; H. Rheinfelder, *Lit. blatt germ. rom. Phil.* 1940, pp. 268–70; R. Glasser, *Rom. Forsch.* 52 (1938), pp. 325–32. This very careful study, based on observations of a stylistic and geographical order, traces the expansion of *diurnum* in literary time. For the south (including Italy) the author can illustrate the advance of *jorn* at the expense of *di* (*dia*) as a victory of Provençal troubadour language. In northern France, however, the struggle between *dies* and *diurnum* had been decided in favour of the latter before the most ancient French texts were written.

64 *Renaut de Montauban* (ed. Michelant, Stuttgart, 1862).

65 *La Mort Aymeri de Narbonne* (ed. Du Parc, Paris, 1884).

66 *La Mort de Garin le Lorrain* (ed. Du Méril, Paris, 1846).

67 *Petronii Cena Trimalchionis* (ed. Heraeus), 41, 10.

68 *Yvain*, v. 5871.

69 *Ibid.*, v. 186.

70 Ed. Tobler, third edition, 1912.

71 Schultz–Gora, *Zwei altfranzös. Dichtungen* (second edition, 1911).

72 E. R. Curtius, *Li Quatre Livre des Reis* (1911).

2
Time in the Middle Ages

Time and eternity

The Middle Ages viewed time from the scientific and the
theoretical standpoints only to an inconsiderable extent and not
at all from that of psychology. The nature of time was a question
that was not posed within these terms of reference. The duty
imposed on medieval humanity by time was one of metaphysical
conformity: 'What is its significance and place in the vast
structure and plan that is the work of God?' In this world, where
all things and values had their place—that is to say, where a high
degree of order prevailed—time was viewed first and foremost in
its relationship to this order. Thought, torn as it was between
Earth and Heaven, the Here and the Beyond, Man and God, was,
as far as the idea of time was concerned, at one with the antithesis of the temporal and the eternal. The difference, or rather
the gulf, separating the human and the divine conceptions of
time is an old biblical and Platonic tradition. The greater the
distance between creator and creature, the less their 'times'
have in common. The more elevated and abstract the conception
of God is, the more strongly man feels his inability to conceive of
God in the light of his earthly time.

This gulf is not the result of logical consequence but the
expression of the religious experience of eternity, a necessary
adjunct of religious piety. A conception of time that was practical
and elaborated to meet the exigencies of life here on earth stood
in opposition to one which had to conform to the idea of the
eternity of God. On this earth, time is divided for the benefit of
man, but in the Beyond temporal divisions no longer play a part.
There, all ordinary notions of duration become meaningless.
Viewed from the Beyond, the reckoning of time appears as
something humanly petty. The abstract, all-powerful and abso-

lute God of Christianity stands outside and above all temporality. Medieval thought, in regarding the world created in time and space as an imperfect manifestation of Being, showed itself to be the heir of ancient philosophy. The idea is immutable and independent of extension and duration. Change and happening detract from its value. To those who hold this view, perceptible reality appears merely as the swinging of a pendulum through a centre of gravity.

In his lucid and profound reflections on time, Saint Augustine decisively drew a line of demarcation between *Deus* and *tempora*. Although he did not presume to define the nature of time ('Si nemo ex me quaerat, scio; si quaerenti explicare velim, nescio'[1]), he was nevertheless aware that the essential form in which it manifests itself is flux. What endures is outside time. And, since God timelessly *is*, there is for Him neither past, nor present, nor future. He dwells in an eternal present. Time itself is a creation of God. The question 'What did God do in the "time" before He created the universe?' is therefore not merely meaningless but blasphemous.[2] Time was created only for man. The latter may be defined as a rational animal subject to time. It thus comes about that *tempus* is never used of the Beyond, and that it is coterminous with 'human' and 'mundane'. A spirit of Purgatory, Marco Lombardo, addressed Dante in these terms:

Or tu chi se' che il nostro fummo fendi,
E di noi parli pur come se tue
Partissi ancor lo tempo per calendi?   [*Purgatorio*, XVI, vv. 25–7]

The essential difference between divine and sublunary things explains why the Beyond is not to be regarded simply as a continuation and a prolongation of earthly existence. Eternity is not a product of multiplication: 'Aeternitas Dei non est tempus infinitum, sed nunc stans.' As with all irrational quantities, it is possible only to say what it is not. The most that can be done is to adumbrate its dissimilarity from the time of man by means of images. Eternity is a day with no yesterday and no tomorrow. The concepts of *stare*, on the one hand, and *transire*, *praeterire* on the other, which Augustine and others after him employed in connection with the Here and the Beyond, are images. The 'aeternitas Dei' resembles the focal point of a lens, in which the

c

spread-out succession of things as they appear to man are concentrated. On earth time is divided, forming a succession, but the eye of God embraces it as an indivisible unity. The scholastic philosopher says of divine existence that

> est sine partibus et sine successione, inde nec in horas, nec dies, nec annos, nec saecula, nec saeculorum myriades describi potest; non est tempus infinitum; semper est una eademque.[3]

Eternity is void not only of external temporal elements, which form the framework of life on earth, but also of emotionally tinged glances directed backwards and forwards in time—remembrance of the past, desire, fear and hope concerning the future—without which we can have no real sense of time. Eternity knows neither the rhythm of the days nor the fluctuating mental states between desire and fulfilment, joy and grief. The circumstance which most strikingly brings out the difference between the human and the divine conceptions of time is the ability of God, stemming necessarily from His omniscience, to see into the future, a power shared to a limited extent by souls in Paradise, who descry on the brow of God earthly events which have not yet taken place. For God, the past and the future have the same degree of reality as the present. But the past is never something over and done with, not only in the eyes of God, but also as far as human destinies are concerned. After death, man awaits the judgment of eternity on his past life, which moreover is present in its entirety and indeed as something more real than a personal recollection. The saying 'only once is never' does not hold good here, because in this case it is not a matter of the perpetuation of the effects of a circumstance but of its existence *per se* and of its inscription in the annals of time.

Between these two poles of the medieval consciousness of time there were, in addition, particular 'times' which the schoolmen considered necessary for purposes of dogma. A distinction was drawn between the time of the physical world and that of pure spirits, whose specific qualities were discriminated by Saint Thomas. It was necessary to avoid confusion between the everlastingness of created entities—the *aevum* of the stars, pure spirits, visionaries and souls in heaven—and the eternity of God. At the same time, other forms of duration were defined, and an

attempt was made to determine their qualitative differences.[4]
Men of learning took such differences very seriously and engaged
in all manner of controversy concerning them. But the general
consciousness of time and the poetic temperament were familiar
only with the tensions and differences of mood between time and
eternity.

Before God, time is nothing. True Christianity is too inward to
attach more than a practical value to temporal quantities. An
entire lifetime of sin may be redeemed by a moment of contri-
tion. Service in the cause of religion was judged according to the
intensity of its devotion, and not by the length of time involved.
Nor does God's mercy stand in any temporal relationship to the
sinful actions of man:

> Magna Dei clementia, quod tam magnum peccatum, cui poeni-
> tentia debetur quindecim annorum, tam brevis delet contritio.[5]

If the elevation of God over time was held to be the highest
degree of being, this was no ground for neglecting or despising
time in everyday life. The proper use and shaping of time was
the road to eternity. But, despite this, there was no question of
estimating its value as an experience. There was no thought of
holding fast to the moment. Duration was not a thing that was
valued in itself. To divide it and adapt oneself to its flow was all
that was 'done with' it. Man had no demands to make on time;
rather, he was beholden to it, and had to adjust himself to it as
part of a divinely ordained scheme of things. For someone to say
that the day or night was too long for him was an expression of
impatience out of keeping with the age; for medieval humanity,
this would have been to pit oneself against time, whose course
was determined by the Creator. To do so was impious. Impatience
was directed only against those temporal circumstances which
lay in its power to modify. Unwillingness to adapt oneself to
God's time, and the desire to make one's own joy or sorrow the
measure of time, was evidence of a mundane spirit.

Such inviolable intervals of time included the monastic *horae*
—laid down by man, it is true, yet dedicated to the service of God.
Punctuality was a virtue that was rigorously insisted on. It was in
keeping with the dutiful character of the medieval monk. Dutiful-
ness showed itself also in the subordination of the individual

to an established temporal order. Late arrival on the part of monks at divine service or at meals was punished. This was not the expression of an enhanced estimation of the value of time, nor was there any thought of 'saving' it, but 'strict regulation of time imparted measure and security to monastic life, and guarded against dissolution and fragmentation'.[6]

The monk's entire life span, and his individual day, were alike regulated by time. To measure it appeared to medieval humanity as one of the most important obligations towards time, because it possessed, apart from its practical utility, symbolic value, and offered opportunity for pious ritual and commemoration. These were of greater import. The seasons of the year were distinguished less by what they offered and brought than by their significance. Day and night were differentiated by diversity of symbolic meaning. The evaluation of a juncture in time as something that occurred but once and having value as an experience was unknown. The Church was the first power to enjoin the measurement of time and a temporally ordered mode of life, and this end it vigorously pursued. The monastic life was a rigorous implementation of the principle of order. Saint Benedict recommended the wearing of a temporal straitjacket which constricted the life of the individual and the subjective sense of time.[7] The strict observation of set times was characteristic of the monkhood to such an extent that the author of *Gargantua* pilloried the clock as a symbol of medieval monastic life.

The desire for the precise temporal regulation of the Christian day—not only in its mundane but also in its spiritual aspects— led, in the year 1000, to the invention by Pope Sylvester II of the striking and wheeled clock. 'The first public clocks appeared in Germany about 1200 and the first pocket watches a little later. The significant association of time measurement with the structure of religious worship should be noted,' Spengler writes.[8] This relationship between the Church and time is indeed noteworthy, but the Church did not, by virtue of its ordinances concerning time, make itself the advocate of a new temporal sense, as Spengler believed. The clock was an instrument of discipline and a means of adapting oneself to something supra-individual. It had its practical function in the service of a higher purpose. What Spengler says of it is therefore false: 'Of all the peoples of the West, it was the Germans who invented the mechanical clock, that grisly symbol of ever-flowing time, whose strokes, resounding day and night from countless clock-towers all over Europe, are perhaps the most monstrous expression of which an historical world-sense is capable.'[9] Spengler, a romantic, confuses the strictly factual significance and purpose of the clock as an instrument of practical time measurement with the effect exerted by countless pealings of bells on his inner auditory sense. By means of this interpretation he attempts to substantiate the difference, alleged by him to exist, between the modern consciousness of time and the absence of it in days of antiquity.

Werner Gent speaks in rather more restrained terms of the significance of the clock: 'Might it be that the invention of the wheeled clock, which took place about the year 1000, fostered the development of a true sense of history?'[10] It is indeed probable that the clock, thanks to the enhanced awareness of time and of temporal relations which it engendered, did prepare the ground for the growth of a new faculty of historical vision. But the clock was not in itself the expression of an historical sense; nor did the Middle Ages regard the clock as such. The striking of the clock, the vehicle for the audible communication of time, was the symbol of a communal spirit. Time summoned everyone without distinction, for it was the affair of all.

The most effective argument against the alleged absence of a sense of time in antiquity is that arbitrary subdivision of the day, the *hora*, which the Middle Ages inherited from antiquity. With its demand for the strict observance of time, the Church took a stand against the popular sense of time, youthful and insouciant, which we discerned in the national epic. Such notions of temporal division as existed in the popular tongue came in the main from the intellectual heritage of the Church. One such concept was that of the hour. As we have seen, the day in that period of history was less a measure of time, less a quantitative unit with subdivisions, than a unit of experience, of light and rhythm. But the swiftly flying day was divided by the Church into hours, *horae*.

The idea of 'the hour' had long existed before it arrived at its colourless, precise, objective meaning. The Greek ὥρα was unknown to Herodotus in the sense of 'hour'. From its general meaning of 'time' it went on to acquire the particular one of 'time of day'. From this, in turn, it took on the meaning of 'hour'. It appears that it was the astronomers who first invested it with this significance; at all events, Hipparchus, as early as 140 B.C., frequently used it in this sense, as the *Almagest* bears witness. The Romans acquired the word along with the sundial.[11] *Hora*, having a fixed, precise meaning, was not popularly used. The word passed with the spread of Christianity into European languages, but these did not associate any precise and objective division of the day with it. In any case there is no question of its having been used in the exact sense with which we associate the word today. Virtually throughout the Middle Ages, the hour was an indeterminate period, varying according to the length of the day.[12]

In ancient French literature the word was scarcely used as an indication of time, such indications being in any event very sparse. One looks in vain for such phrases as *à trois heures, une demi-heure, à toutes les heures du jour*, and so on. The grammarian who is aware of the existence of *heure* and *jour* and is familiar with their syntactical function in Old French might consider such a verbal combination as *à toutes les heures du jour* to be possible from his own standpoint. But in fact there was no connection between the two concepts in the ancient tongue. They had nothing to do with one another. The hour was not as yet thought of as a part of the day; in the popular tongue it had, initially, no definite, more or less precise durational significance, but was in general the equivalent of 'point in time'. It was not used as a unit of duration before the time of Middle French, being used relatively to the day and other periods:

Ja n'el servisses un esté,
Non pas un jor, non pas une hore.   [*Roman de la Rose*, v. 4862]

In Old French the word *hora* did not undergo its linguistic development, properly so called, as an indication of length of time but as an imprecise temporal index used in conjunction with a number of newly formed adverbs: *or, ores, encore, désormais, dorénavant, orendroit, alors, d'ores et déjà* and *lorsque*. These new terms entirely obscured the concept of 'hour', for which the language initially found little use. The linguistic development of this word in an age where we find in the popular tongue so little evidence of some measure of desire and ability for the objective expression of time is reflected in the insufficient attention paid to the dating of even Latin texts. 'Il avait été prescrit par la législation romaine que tout acte, pour avoir une valeur, devait être daté de l'année et du jour, et cette prescription avait passé dans plusieurs lois barbares. Cependant nombre de documents de toutes les époques, mais particulièrement du IXe au XIIe siècle, ne sont pas datés ou n'ont qu'une date insuffisante'.[13] The reckoning of time had gone by the board, so that a familiarity with this science was one of the most important elements of medieval cultural activity. The reckoning of time, however, was used less for the dating of historical events than for the calculation of ecclesiastical feast days. This function invested it, in the eyes of the age, with its value and attractiveness. Thus, for instance, the *Cumpoz* of the Norman poet Philippe de Thaun[14] served this pious purpose; the work raised instruction to a devotional level. The temporal concepts of the Julian calendar were represented with symbolic and etymological explanations. His interpretation of temporal concepts, and his investing them with a religious element, lent them their value in his own eyes and those of others. There was no abstract reckoning of time with measures and periods that were devoid of content.

By dint of this work of elucidation the popular tongue underwent a conceptual enrichment, even if all the terms to be added to the French language were not introduced by Philippe de Thaun. Additions to the spoken tongue included the following: *anuel* (50), *atomete* (2321), *autumnal* (3238), *bissexte* (189), *embolismaisun* (2310), *embolisme* (192), *epactes* (198), *equinoctiun* (199), *feries* (487), *estival* (3255), *hivernal* (3256), *hurete* (2322), *ides* (185), *indiction* (202), *kalendes* (185), *kalendier* (2493), *lunaisun* (2292), *lunal* (797), *momenz* (2321), *none* (186), *prime* (251), *quinte* (253), *salt* (2346), *semaine* (409), *siste* (254), *solsticiun* (3253), *terme* (201), *tierce* (252), *ver* ('spring') (1918), *vernal* (2265), *zodiacus* (397) and, in addition, the names of the months and the days of the week. The terms listed above were only the scaffolding for the division, ordering and reckoning of the year. Ides and kalends, phases of the moon, the zodiac, equinoxes and solstices were no doubt necessary elements of learning, but provided no stimulus to the imagination and possessed no intrinsic value. As constituent elements of the natural course of the year, they receded before the year that stood on a higher plane, which in the mind of the believer ousted the natural year. Time in the Middle Ages was neither an historical nor a natural process. The natural elements of time, such as the year and human life, were raised to an intellectual and symbolic status. The calendar, with its many feasts, provided an ever-recurring series of spiritual oases throughout the year. It possessed a higher degree of reality than the natural year, with its succession of seasons. If Jacob de Voragine based his *Golden Legend* neither on the succession of saints in factual history nor on any other material, and if the succession of feasts in the ecclesiastical year is the foundation for his collection of stories, it was because he considered this succession to afford the most apt and elevated basis for its compilation.[15]

Temporal formalism

Life and its course were a process in taking stock of which humanity looked beyond its tangible and mutable circumstances in an endeavour to discern an abstract, formal and meaningful framework. There was an ideal life span of fourscore years and ten, there were four ages of life, which in their succession and significance were equated with the four periods of the day and the four seasons of the year. Since time had a meaning, the time of the death of Christ must also be meaningful. The high-water mark of life was the age of thirty-five years. Christ met His death at the age of thirty-three, for it was not fitting for Him to live through the declining years of life.[16] But the hour of His death was significant also, for the sixth hour was the zenith of the day. Thus Jesus died at the very prime of life, since He, as Dante observed, was *ottimamente naturato*.[17] In this view of time, no place was given to chance. Events—particularly scriptural events —had necessarily to occur within these subsequently recognized temporal premises. Several considerations were bound up with the fact that Christ was resurrected on the third day after His crucifixion, considerations which invested this span of time with its symbolic value: (1) The day during which the Saviour lay in His grave signified the light of His death; the two nights, our twofold death. (2) The three days bore witness to the truth of these happenings. (3) This period of time was an expression of the power of Christ. (4) The three days symbolized everything brought about by Christ, in Heaven, on Earth and in Hell. (5) This span of time was the expression of the threefold lot of the just: suffering, rest and joy.[18]

We adduce this example in order to show how medieval thought revolved round and clung to such details, which appear to us to be external and incidental. In medieval thought an indication of time which in our estimation might not have corresponded to the facts retained its inviolability and weight and became wide-ranging in its scope. The interval between Good Friday and Easter Sunday was an ever-present reality encountered at every turn by anyone with an open mind for such circumstances. Thus we see that in many medieval indications of time there is an implied adumbration of mysterious associations and meanings. If one reads the numerous observations of Jacob de Voragine on the temporal circumstances mentioned by him, one realizes that every indication of time was firmly anchored in the imagination by associations shared by all. The medieval interpretation of process was not the outcome of knowledge gained from experience but the result of a mental attitude which idealized isolated cases handed down by tradition. There were ideal intervals of time, and the medieval mind identified such intervals with those exemplary instances of salutary change or divine dispensation which it perceived to take place.

Biblical history unrolled itself according to an ideal temporal plan. When we read in *The Golden Legend* that God sent His son among men 'at an opportune time', this must not be understood as meaning at an opportune time in the historical sense. The significance of the phrase is that a temporal scheme ordained by God had arrived at its appointed end. Thus the phrase used by Jacob de Voragine must not be understood as meaning that the coming of Christ occurred at an opportune time only from the viewpoint of humanity, but rather that 'it was also the opportune time in the temporal sense, for it is written "when the fulness of time was come" (Galatians, 4, iv).'[19] Time had its own laws, that is to say, its course seemed inexorable when set against humanity and its will and desires. Events were subordinated to their constituent elements; there was no overlapping from one element to another. It was not the events of life that formed periods and intervals in time; these periods and intervals existed before the events—the frame existed before the picture. Time did not flow like a liquid but marched on in strides of unvarying length.

The normal stride of time was a day. In popular medieval thought the day was by far the most important unit of time. The next most important unit was the year. All others—hours, weeks, months—were relegated to a much lower place. Many successive individual happenings that went to make up a more general event were 'stylized' into occurrences of equal length by virtue of their being brought within the framework of a single day. It must not, however, be concluded from this that every event was thought of as having literally the same duration. Each event was, as it were, brought into line with the characteristics of a particular day and hence became inseparable from it, so that the tenor of an event corresponded with that of the day. The archetype of this 'stylization' was the biblical history of the Creation. In the Middle Ages great use was made of the possibilities inherent in such conceits.

Let us take as an example the manner in which the omens of the Last Judgment were imagined as successively presenting themselves:

> On the first day, the sea will rise to a height of forty ells, covering every mountain and standing as a wall in its appointed place. On the second day, the sea will recede so greatly that it will scarce be able to be seen. On the third day, the sea will rise again and will raise its voice against heaven. Its voice will be heard by God alone. On the fourth day, the sea and all the waters will stand in flames. On the fifth day, every tree and wort will give forth a dew of bloody hue. It is said also, that all the birds of the air will gather and will dare not either eat or drink from fear of the coming of the stern Judge. On the sixth day, the cities will fall asunder and all that is builded, and lightning will flash across the face of the firmament from the setting of the sun until its rising. On the seventh day, all the stones will smite themselves against one another so that they

will be sundered, and each will be cleft into four parts, whereof each will rub itself against its neighbour. Of the clamour that will arise, no one will have knowledge save God alone. On the eighth day, a mighty earthquake will come about, so great that all men and beasts will fall to the ground and will be powerless to stand on their feet. On the ninth day, everything on the face of the earth will be levelled, and all mountains and hills will be sundered to dust. On the tenth day, men will come forth from the caves whither they have fled as though possessed of the devil and speechless. On the eleventh day, the bones of the dead will arise and stand over their graves. And all graves will be opened from the setting of the sun until its rising, so that the dead may come forth. On the twelfth day, the stars will fall from the heavens and all planets and fixed stars will give forth sheaves of fire; and again there will fall a rain of fire. It is said also, that on that day all beasts of the fields will assemble amid a great bellowing and will eat and drink not. On the thirteenth day, the living will die and arise with the dead. On the fourteenth day, heaven and earth will stand in flames. On the fifteenth day, a new heaven and earth will come into being, and all men will arise again.　　　　　[*Leg. Aur.*, I, col. 9–10]

Temporal formalism played an important part where events were not yet regarded from the point of view of science and historical development, and where simple causal connections were not understood. There was a tendency to interpret a fact— a sick man being cured in seven days, for instance—in the light of temporal symbolism. Everything that happened came about according to mysterious temporal dispensations. Religious thought was coloured to a marked extent by such fancies. Even the entire medieval world fell into the categories of the holy and the profane, of good and evil, so there was for the Christian the hour of God and the hour of the devil. Time was not something neutral but a force either sinister or beneficent which imparted its character to a circumstance. In the eyes of this mode of thought, not only was a deed itself evil, but also the time of its conception or execution.

The hour in which evil came about was held to be jointly responsible for such evil. In the light of this attitude, one may readily understand the frequent use made in the Middle Ages of exclamations such as *maleoite soit l'ore*. In the medieval mind the happenings of a particular day were intimately linked with the day itself. Viewed from this standpoint, commemorative dates and the return of eventful days acquired a particular significance. It was believed also that the same calendar days of different years were in some manner bound up with one another. Indeed, their relationship might almost be described as one of identity. They corresponded with one another both in their inner nature and in their effect. What the first of May brought with it in one year might be annulled in the following twelve-month. A legend relates how a monk of the monastery of Michaelstein noticed that his knowledge of Latin vanished utterly from his mind concurrently with a loss of blood he had suffered, with the result that he identified the one with the other. That the relationship might have been one of cause and effect never occurred to him. On the advice of another, he opened a vein precisely a year afterwards ('eodem die et eadem hora'), thereby regaining his lost knowledge.[20] In the tale of the *Chevalier au Barisel*[21] the period of atonement was one year from Good Friday to Good Friday. In order that the cask of the sinner might be filled, not only was painful remorse required but the formal, implicit condition of the elapsing of a span of time was necessary also. To those who held this conception of time, haste and impatience must have seemed senseless. The future was, in some mysterious way, already contained in the present. What happened today was the immediate cause of temporal effects at a distance, which would manifest themselves a year or a hundred years hence at the same hour. At the root of this lay the idea that the future appeared as such only to mankind, whereas in the metaphysical sense it already existed.

In the light of this conception, time had no concrete reality; it did not change things slowly and gradually, no becoming and flux was observed; one did not become conscious of time as the content of events, but as a point-like entity. It lacked the fluctuating and gliding movement of continuity; what happened now required no immediate continuation in the next moment or the next day, but was in a state of repose, with no connection with any other time except a point in the future linked to the present by an invisible bridge spanning all intervening events.

This relationship between the present and the future did not, however, become apparent to the medieval mind when the future had, in turn, become the present. This is borne out by the case of the monk who forfeited his knowledge of Latin. A causal link between the past and the present was, nevertheless, often established. An individual, as he cursed the hour of his birth, saw the root of present evil in a remote point in the past. This point in the past was not denounced as a melancholy and evil circumstance in itself, but only in relation to its dire effects in the future. Given this attitude towards time, what was over was not done with. The hour of one's birth was ever-present, in a sense other than the astrological, the biological or the historical one.

Temporal formalism meant also that a supernatural construction was placed on temporal coincidences. About the middle of the fifteenth century it was noticed in Ghent that the body of Saint Bertulf, buried in an abbey there, was stirring violently.[22] These movements precisely coincided in time with an earthquake then being experienced. At this same time the duke of Burgundy was due to enter the city with the Dauphin. This combination of circumstances was interpreted as an evil omen. A hundred other like examples might be quoted. In the consciousness of those centuries, time possessed a peculiar power to bind; everything that happened at the same moment was connected by a mysterious link. Even those things which were causally and spatially remote were related by virtue of temporal coincidence. Accordingly, time was thought of as a medium in which divine power or daemonic machinations were reflected. Simultaneity was an index, a signpost for the direction which pious thought should take. In *The Golden Legend* it is reported how a God-fearing man heard heavenly music on 8th September every year, and how he asked of God that He might reveal to him the significance of these sounds. He was told that it was the anniversary of the birth of Mary. In this way not only was a calendar with significant dates from the point of view of religious history offered and proclaimed to mankind, but the faith of humanity was itself the source of new commemorative days.

Such a conception of time naturally had its influence on human conduct. It inculcated, in fact, the idea that one was obliged as a Christian to lead a very austere mode of life, in which subjective desires were stringently curbed. On holy days no profane act might be performed, nor anything which might not be in keeping with their solemnity. It was, for instance, a crime to bathe on the night of the birth of Mary, and accordingly the Christian sense of rectitude demanded that anyone disobeying this interdiction should be put to death.[23] Thus we see that, in this respect, the medieval mind was far from holding a natural and flexible view of time. Time was not geared to circumstances and events but possessed a monumental fixity, for it had its own independent existence. One believed in it and complied with its demands all the more in proportion as one refrained from speculating as to its nature. This time that lay outside experience imparted to the face of the Middle Ages an air of abstraction and dissociation from reality—qualities embodied in the Gothic cathedral.

Just as periods of time were symbolically evaluated, so did the pace and speed of a process appear as the manifestation of some extra-terrestrial mind and force. Thus lightning-like rapidity suggested divine associations or origin. God was master of time in every respect, and not the least proof of His mastery was the fact that His acts were independent of time. He possessed the power of making things happen with the celerity of lightning. With Dante, speed was sometimes a symbol of divine, sometimes of diabolical power, now infinitely exalted, now infinitely sinister. The proximity of the heavenly spheres to the Deity in the *Paradiso* is reflected in the varying degrees of rapidity with which they revolve.

Temporal formalism towered above religious life and thought and—proof of its degree of immanence in the medieval mind— played an important part in the worldly domain of chivalry. The hero Siegfried sojourned an entire year in Worms before he made his first encounter with Kriemhild, on whose behalf he had come. The spirit of chivalry intercalated a lengthy span of time between the conception of a wish and its fulfilment, and stylized the natural course of events by subordinating them to the conditions of temporal formalism. The courtly upbringing of the knight was essentially an objectivization and a canalization of existing spiritual and physical forces. The knight was required to perform objective and unequivocal tasks, objective touchstones of his capabilities. Amongst these was numbered the observance of a fixed period of time. He was set a definite period of time to which he was bound to adhere unconditionally and precisely.

This is a motif which played a great role in popular tales.[24]

In the courtly romance the internal and external course of the narrative reposed on the awareness of the need to observe an arbitrarily established but irrevocable time span. In these romances time occupied a more prominent place than it did in the *chanson de geste*. In *Yvain* the set period was a predominant element. We repeatedly observe it to influence the Knight's fate. When he was prevailed upon by Gauvain to embark on an adventure in his company, so that he might not idle away his time, his wife Laudine granted him a year's leave of absence.[25] During this time she sat at home and counted the days 'that came and went'[26] on a calendar in her room. The fact that the hero failed to return on the expiration of the appointed period determined his fate during the second half of the story. The same motif reappears in other episodes. His rescuer, Lunete, was granted a period of forty days in which to find a knight who would champion her innocence in combat.[27] King Arthur like-wise granted a span of forty days to one of the two maidens who were disputing their share in the inheritance, during which time she acquired the services of Yvain in defence of her cause.[28] In the execution of his heroic deeds Yvain had to observe certain stipulations in the matter of time. He was allowed to fight the giant on behalf of Gauvain's niece only on condition that the giant should present himself for the combat before the following morning was over.[29]

Thus we see that in the work of Chrestien de Troyes the decision on important matters often hinged on the question of time. Punctual arrival was frequently a circumstance which inclined things to turn out for the best. Much depended on a single day. Let us consider the following conversation between Laudine and Lunete:

'Dame! Ne cuit que nus oisiaus
Poïst en un jor tant voler.
Mes je i ferai ja aler
Un mien garçon, qui mout tost cort,
Qui ira bien jusqu'a la cort
Le roi Artu au mien espoir
Au mains jusqu'a demain au soir;
Que jusque la n'iert il trovez.'
'Cist termes est trop lons assez.

Li jor sont lonc. Mes dites li,
Que demain au soir resoit ci
Et aut plus tost, que il ne siaut;
Car, se bien esforcier se viaut,
Fera de deus jornees une.
Et anquenuit luira la lune,
Si reface de la nuit jor.
Et je li donrai au retor,
Quanqu'il voldra, que je li doigne.'   [vv. 1824–41]

Here the difference between the national epic and the courtly
romance becomes apparent in so far as their treatment of time is
concerned. In the *Chanson de Roland* the arrival of Charlemagne
was not significantly bound up with time. In contrast to this,
Chrestien's gaze was firmly fixed on the point in time on which
the decision depended. The poet dwelt on it with the intention of
heightening the expectancy of his readers or audience by keeping
uppermost in their minds the possibility of a belated arrival; this
means of inducing tension was also a feature of the popular tale.
To Yvain, who arrived in the nick of time, thus preserving
Lunete from death, she said:

'S'un po eüssiez plus esté,
Par tans fusse charbons et çandre.'   [vv. 4406–7]

In the work of Chrestien de Troyes this regard for time appears
first and foremost as an ethical duty. It may, however, be con-
strued as an aesthetic formulation. Every manifestation of life
that concerns itself with beauty makes of time a means of em-
bellishing art or life. In the life of the knight such manifestations
were stylized in the form of the set period of time. The world in
which he lived was stabilized and ennobled by means of the
temporal obligations to which he was bound; the world, from
something naturally given, was transmuted into something
artistic. This sense of temporal propriety, which curbed natural
impulses and transformed the involuntary into the voluntary,
was always linked with the idea of a refined and noble mode of
life.

A prerequisite for the assessment of the essential nature of such
temporal demands, of a custom, of a temporally formulated
licence or interdiction, was an awareness of whether they were
relevant to time or not. Herein is contained the distinction we

shall exemplify in the following circumstance: in order to prevent a marriage feast from degenerating into an unbridled carousal the authorities in former centuries decreed that such a feast might not be celebrated for more than one day. This disposition was without true temporal relevance, for the prescribed period had no value or significance in itself. The interdiction might take a different, non-temporal, form and still fulfil its purpose: for example, it might be ordered that only so many glasses of beer or wine might be drunk. Another instance which doubtless falls into the same category is the examination of candidates for a degree, which lasts one hour in their main subjects and half an hour in each of their subsidiary ones. The length of time during which the students are examined is essentially unimportant, for it is not really a question of time but of the extent of the knowledge they are required to display. In this case, therefore, the stipulations with regard to time serve a practical purpose outside the sphere of temporal values. They have, that is to say, no relevance to time. Here, time is only an instrument to be used. In contrast to this stood the three days, the three years, the seven years, the hundred years allotted to the hero of the popular tale— periods universally acknowledged by the medieval mentality, which set much store by symbolism. Here, temporal considerations had a true relevance. The observance or otherwise of the set period was strictly a matter of time, which had its own beauty and value. That Sleeping Beauty had to sleep for a hundred years was accounted for by nothing in the nature of things outside time. Only the law of time was valid, a particular aesthetic unaffected by practical and historical considerations.

But in the realities of life—medieval life not excluded—these two modes of temporal determination were never met with in a pure state, that is to say, the one never completely excluded the other. Stylization, which was relevant to time, and ex-pediency, which was not, are difficult to distinguish from one another as mainsprings of temporal demands. Medieval thought nonetheless attempted to suppress the temporally irrelevant in as wide a sphere as possible. The temporally irrelevant, however, played an increasingly important part in the temporal forms of life and art in proportion as pure civilization advanced. In the medieval consciousness, time as a quantity was an autonomously existing entity with its own value, derived from no other source; time, therefore, was fitted to lend value to other things.

This value resided in the conception, peculiar to that age, of the temporal unit. A year was not considered as being merely the sum of three hundred and sixty-five days, or of twelve months, but as an organic unit which would be destroyed by the removal of its smallest part. If a year lacked a single day it was valueless. If a twelvemonth, set as a fixed period, was overstepped by an hour, it lost its significance. In this view of time there were no approximations. Here we have the explanation of the high seriousness with which temporal demands, of obscure practical expediency, were regarded. The ancient belief in the connection, familiar to us from old popular tales, becomes apparent here. Time was the instrument of an uncanny will which set itself against mankind. Even the period which the *châtelaine* required the knight to observe had all the intangibility of a supernatural demand. The one who prescribed the set period, no less than the one who had to observe it, was bound to it without appeal, for the former was as little master of time as the latter. When it was a question of time which lay at one's own disposal, scant regard was paid to it. No individual might lay claim to time for the sake of his personal advantage. The strict observance in the Middle Ages of temporal impositions and dispositions was not a case of consideration for others in the matter of time, but the acknowledgement of a higher authority.

Notes

1 *Confess.*, XI, 14.
2 Cf. Kurt Reichenberger, *Die Schöpfungswoche des Du Bartas* (Tübingen, 1963), II: *Themen und Quellen der Sepmaine*, p. 22.
3 *Institutio Philosophiae Wolfianae*, 526 (1735).
4 Friedrich Beemelmans, *Zeit und Ewigkeit nach Thomas von Aquino* (Münster, 1914).
5 Caes. Heisterbach, *Dial. Mir.* (Cologne, 1851), II, 11.
6 P. Hoffmann, *Der mittelalterliche Mensch, gesehen aus Welt und Umwelt Notkers des Deutschen* (Gotha, 1922), p. 83.
7 *S. Benedicti Regula Monasteriorum* (ed. Linderbauer, Bonn, 1928), Nos. 8–17 of the rule.
8 *Der Untergang des Abendlandes*, I, p. 19, n. 1.

9 *Op. cit.*, I, p. 18.
10 W. Gent, *op. cit.*, p. 155.
11 L. Ideler, *Handbuch der mathematischen und technischen Chronologie* (Berlin, 1825), I, p. 238.
12 Nilsson, *op. cit.*, pp. 43, 44.
13 A. Giry, *Manuel de Diplomatique* (Paris, 1894), p. 577.
14 *Li Cumpoz Philipe de Thaun* (ed. Mall, Strasbourg, 1873).
15 J. de Voragine, *Legenda Aurea* (German version by R. Benz, Jena, 1925), p. xiii.
16 Dante, *Il Convivio*, IV, chap. 23.
17 *Ibid.*, IV, chap. 23.
18 *Leg. Aur.* (Benz), I, col. 354–6.
19 *Ibid.*, I, col. 4.
20 Caes. Heisterback, *Dial. Mir.* X, 4.
21 *Zwei altfranzös. Dichtungen* (ed. Schultz–Gora).
22 Chastellain, III, pp. 407, 408.
23 *Leg. Aur.* (Benz), II, col. 132.
24 Cf. R. Glasser, 'Der Zeitbegriff des Märchens', *Geistige Arbeit* (1936), No. 23. On the part played by time in the fairy-tale cf. Hedwig von Beit, *Das Märchen: sein Ort in der geistigen Entwicklung* (Berne and Munich, 1965), *passim*. According to von Beit, 'the exact indication of a point in, or period of, time resembles a foreign body which has insinuated itself into archaic space'.
25 *Yvain*, vv. 2570f.
26 *Ibid.*, v. 2795.
27 *Ibid.*, v. 3691.
28 *Ibid.*, v. 4803.
29 *Ibid.*, vv. 3944f.

3

The concept of time in the later Middle Ages

Objective time

There is a contradiction between what has been said concerning the ancient French view of time and the extended primacy of temporally formalistic thought in the Middle Ages. On the one side we observe a striking unconcern for temporal things, and on the other an attitude of absolute seriousness towards time and its relationships. In the earlier age, years and centuries went unheeded; in the later, instants were treated with reverence. This antithesis may be explained by the dual aspect of the same attitude of mind. If we have spoken of a lack of concern for temporal relations, this bald statement calls for some qualification. In point of fact, it was only with regard to the temporal relationships of reality in their scientific and historical aspects that interest and insight were lacking. That time in which earthly things came and went was immaterial. Time did indeed attract the attention of the medieval mentality when the latter believed that it discerned the manifestation of a supernatural power in the shape of a temporal circumstance.

Gradually, however, this attitude towards time changed. In the later centuries of the Middle Ages humanity became increasingly conscious of earthly time and of its value and significance for the life here below. With this heightened awareness, the former indifference disappeared. One was in the world and one had to come to terms with time. The sense of temporal order increased. It became a source of recreation to date important developments in daily life and in the cultural sphere. The practice was indulged in even in the realm of poetry.[1] It could, indeed, happen that dating did not serve simply to set the content of a piece in its temporal context, but that dating became an end in itself, a species of ornament added to the text. On the occasion of the birth of Charles VI and his brother, Louis d'Orléans, Eustache Deschamps composed a ballad, the first three stanzas of which read:

En dimenche, le tiers jour de decembre,
L'an mil CCC avec soixante et huit,
Fut a Saint Pol nez dedens une chambre
Charles li Roys, trois heures puis minuit,
Filz de Charles cinquiesme de ce nom,
Roy des François, de Jehane de Bourbon,
Roine a ce temps couronnee de France,
Le premier jour de l'Advent qui fut bon:
Par ce sçara chascun ceste naissance.

Ou signe estoit, si comme je me membre,
De la Vierge la lune en celle nuit,
En la face seconde; et si me remembre
Qu'au sixte jour dudit mois fut conduit
Et baptizié a Saint Pol, ce scet on,
Ou il avoit maint prince et maint baron;
Montmorancy, Dampmartin sanz doubtance.
Tous deux Charles leverent l'enfançon:
Par ce sçara chascun cette naissance.

Trois ans apres, quant li mois de mars entre
A treize jour, sabmedi, saichent tuit,
L'an mil CCCLX et onze, entendre
Puet un chascun la naissance et le bruit
De Loys né, frere du Roy Charlon,
Apres mie nuit trois heures environ;
La lune estoit a IX jours de croissance;
Marraine fut madame d'Alençon:
Par ce sçara chascun cette naissance.   [I, p. 146]

These versified certificates of birth and baptism were characteristic of an age in which the mind strained to find its bearings in the world of external relations. The eye sought to see clearly; one was not carried along, as before, by the rhythm of events, but one set oneself at a distance from them by ordering and observing them. Time became a system for ordering things, and its value as an experience declined. But a sharper perception developed for what might be called 'temporal apparatus'. A date was, for the mind, not something empty but conjured up those associations linked with the gravity and dignity of the calendar and of the Christian reckoning of time. By virtue of the calendar day on which he was born the individual stood in an all-important relationship with the course of Christian history. But even without this religious association, time had become something perceptible, a visible order whose beauty was appreciated. In the later Middle Ages the aesthetic aspect of the calendar was discovered. For the mentality of those days the clock became a symbol of a well ordered course of events, and a model of a meaningful and pleasing co-ordination. Humanity increasingly fell in with an objective temporal order to which all were subject. Those who failed to adapt themselves to it were ridiculed by their contemporaries. One such was the priest in the *Cent Nouvelles Nouvelles*, who had lost his sense of time to such an extent that it belatedly occurred to him, as he journeyed to the city, that the morrow was Palm Sunday.[2]

Time was a means to the ordering of relations in the rising middle classes. As human intercourse acquired set forms, as general and personal life became more sharply demarcated, and as communal life became more close-knit and integrated social groups were set up, the greater the need for temporal order within the community became. Clocks and calendars became more widespread, and public and private life were regulated by them. The temporal course of all events was observed and reported by men of letters. 'L'élément "temps" devient un facteur précieux, parce qu'on raconte les événements non en fonction d'une idéale épopée, mais en fonction de la vie commune.'[3] In the prose version of *Lancelot* objective time was incorporated, as a circumstance determining the existence of humanity, in the concept of reality that had to be depicted.[4] Happenings were seen in their temporal relationship to others. They were viewed from outside, in their extension in space and time. In the texts of this era we notice a great increase in the temporal attributes of the verb as compared with those obtaining in the age of Old French. Sometimes the narrator's attention was fixed on a succeeding event, and sometimes he looked back on a preceding one from the standpoint of the one he was describing. The sense of temporal coincidence and distance was ever alert. The conception of time was far removed from that of the *Chanson de Roland*. The questions 'when?' and 'how long?' were now constantly asked. There was less imagined personal experience of, and involvement in, circumstances, and appreciation of their interrelationship became enhanced.

The attitude of humanity towards time was not one of subservience and adaptation to a rigid system; instead, time was shaped by a lust for life arising from a sense of mental and physical freedom. Objective time was not resented as a straitjacket. This feeling was first experienced by Rabelais. One was at home in external time. The hour at which an event occurred was stated, and its duration was also frequently expressed in hours. Historians considered precise indications of time as an attribute of serious historical writing. Let us note the exactitude with which the time of day, and duration in terms of hours, were given in their accounts of events:

Monstrelet:[5]
bataille, qui dura trois heures. [I, p. 72]

ils actendirent leurs adversaires bien heure et demie. [I, p. 78]

Commynes:[6]
Mais ce fut grant chose, à mon advis, de se rallier sur le champ et estre trois ou quatre heures en cest estat, l'un devant l'autre.
[I, 4 (I, p. 32)]

promisdrent bailler trois ostaiges... et les livrer dedans le lendemain huyt heures. [II, 1 (I, p. 98)]

Sur l'heure de dix heures du matin se trouvèrent en ung villaige fort. [II, 2 (I, p. 105)]

Comme vindrent les neuf heures au soir, nous ouysmes sonner leur cloche. [II, 3 (I, p. 112)]

... furent long temps à ce palais, jusques à bien deux heures après mynuyt. [II, 3 (I, p. 114)]

Demourèrent hors ladicte ville depuis deux heures après mynuict jusques à six heures du matin. [II, 10 (I, p. 149)]

En moins d'ung quart d'heure chascun portoit ladicte livrée.
[III, 6 (I, p. 209)]

N'y avoit point ung quart d'heure que j'estois arriveé.
[VIII, 11 (III, p. 183)]

Death of Charles VIII:
En disant ceste parolle, cheüt à l'envers et perdit la parolle. Il pouvoit estre deux heures après mydi; et demoura audict lieu jusques à unze heures de nuyt... le trouvoit-l'on couché sur une

pouvre paillasse, dont jamais ne partit jusques il eust rendu l'ame ; et y fut neuf heures.   [VIII, 25 (III, p. 307)]

A further advantage of the use of 'hour' lay in the fact that it supplied universally valid and unmistakable determinations of times and duration in place of subjective temporal indications such as 'shortly before', 'soon afterwards', 'a considerable while', and so on, open to arbitrary and erroneous interpretation. Such determinations, however, were not used in a formalistic sense nor in a *voulu* spirit. They were employed relatively and with regard for the need of exactitude demanded by the circumstance under consideration. They appear meaningful and well chosen.

The modern reader, however, must not be misled into comparing this medieval exactitude with that of the twentieth century. The everyday reality of the great mass of mankind was not regulated by the clock. The latter did not become the constant companion, as it were, of the working man until the present century.[7]

Events were now, so to speak, provided with a label assigning them to their place in time. This need for temporal accuracy evidenced itself in narratives in which historical exactitude was not essential. Thus the author of *Petit Jehan de Saintré* often precisely stated the number of hours which the young lover nightly spent with his 'Dame des belles cousines', as though these indications of time were essential if one were to form an accurate picture of their life together. Once they stayed together until the stroke of midnight, 'dont furent tout esbahys'.[8] On another occasion Saintré remained from eleven in the evening until two hours after midnight in the company of his lady.[9] She subsequently did penance for her sins until two in the afternoon in the presence of the abbot of a nearby monastery.[10] These folk lived in the knowledge that they must regulate their lives by the clock. Sometimes they grew oblivious of time, only to start at the sound of the bell, just as people do today. Time frequently jolted them into breaking off what they were doing, hurrying hither and thither, abruptly parting company and taking rapid action. In this respect they appear as puppets whose movements are controlled from outside. This is a qualification of what has already been said. Their haste was represented not as an inner urge but as a necessity dictated by the clock and the state of things. The objective exigency of making haste was conveyed by new expressions:

Le Testament de Pathelin:
Hélas! Despechez-vous, ma femme:
Il est jà tard, l'heure s'approche.[11]

Roi René:[12]
Ce besoing dont j'ay si hastive necessité [IV, 5]

Chastellain:
trouvèrent la chose si mise en estroit que à peine souffroit-elle
délay. [I, p. 75]

Saintré:
Ma dame, qui oyt sonner mydi, se voult haster de partir.
[Ch. 69]

This enhanced awareness of external time and its demands
resulted in a conceptual enrichment of thought. At no period of
French linguistic history were so many terms connected with
time added to the language as in the later Middle Ages. Old
French suffered from a paucity of such terms. The ability to gain
a clear idea of temporal relations went hand in hand with the
new light thrown on the field of temporal possibilities by means
of new words and phrases. The manner in which these additions
were spread over the centuries is indicated by the following list,
which lays no claim to completeness. It consists predominantly
of terms still in use today.[13]

Twelfth century and earlier:

*abréger, annuel, antan, atermoyer, autumnal, bissexte, calendes,
calendier, embolisme, épacte, équinoxe, estival, éternel, éternité,
féries, final, hivernal, ides, indiction, lunaison, lunal, moment,
none, perennité, permanent, présent, prime, primevoire, quinte,
quotidien, salt, semaine, siste, solsticiun, subit, successeur, tardif,
temporel, terme, tierce, ver;* and the names of the days of the
week and the months.[14]

Thirteenth century:

*anniversaire, antique, automne, conséquence, date, férial, futur,
intervalle, minute, pénultième, préterit, prompt, quartaine,
sempiternel, solstice, suranné.*

Fourteenth century:

annuité, antécédent, antéséquent,[15] *célérité, centenaire, chronique*
('chronic'), *par conséquent,*[16] *décade, éphémère, fréquent,
immédiat, imminent, incontinent, instant* ('urgent'), *interstice,*

> *lustre, matutinal, neuvaine, nocturne, nonagénaire, période, perpétuel, perpétuer, à perpétuité, ponctuel, postérité, posthume, priorité, quadragésime, rare, septuagénaire, solsticial, subséquent, succéder, succession, successivement, transitoire.*[17]

Fifteenth century:

> *accélérer, annal, antépénultième, antérieur, coéternel,*[18] *consécutif, contemporain, diurne, diuturne,*[19] *laps, postérieur, primordial, récent, sexagénaire, subsécutivement,*[20] *temporiser.*

Sixteenth century:

> *annales, biennal, décennal, diaire, diuturnité,*[21] *ère, éterniser,*[22] *hebdomadaire, hégire, immémorial, instant* ('moment'), *intérim, postiche, quadragénaire, quinquagénaire, quintane, rapide, rapidité,*[21] *semestre* ('six-monthly'), *temporaire, trimestre, vétusté.*

Seventeenth century:

> *anachronisme, horaire, semestre* ('half-year'), *vicennal.*

Eighteenth century:

> *abrupt, adventice, annuaire, bisannuel, époque, isochrone, longévité, métachronisme, parachronisme, septennal, simultané, synchrone;* and the terms of the Republican calendar.

In conjunction with these terms, thought was enriched by others which, though they cannot be described as strictly temporal concepts, nevertheless express a temporal relationship. Amongst them are numbered those terms which indicate an action or a circumstance and, in addition, place them in a temporal relationship to others and thus orientate them in time. Whereas *vision* for example, expresses no temporal relationship, *prévision* does imply one. The temporal relationship was expressed by a prefix. There were few such words in Old French. In that tongue the need and capacity to express such relations were but feebly developed. However, the number of temporally orientated words and phrases which penetrated the language over the centuries, partly as direct borrowing from Latin, and partly as new formations in French, leaves us in no doubt as to the extent to which the sense of relations in time developed:

Twelfth century:

> *prédestiner, prescience, présumer, réparer, répéter, résonner, ressusciter, successeur.*

Thirteenth century:

ainzjournée, devant veoir,[23] *interjection, prédécesseur, prédire, préjudice, prémisse, prérogative, prévoir, primogéniture.*

Thirteenth and fourteenth centuries:

préambule, préface.

Fourteenth century:

antécédent, anticiper, intermission, interrègne, intervenir, préalable, précéder, préfix ('predetermined'), *prémunir, préméditation, préoccupation, préparer, présage, prévisif, prévision, récapituler, récollection, réitérer.*

Fifteenth century:

anticipation, antidater, préavertir,[24] *préavisement,*[25] *précognoistre,*[26] *précurseur, préfigurer,*[27] *prémices,*[28] *prénostiquer,*[29] *présagissement,*[30] *présupposer,*[31] *prévenir,*[32] *prévention.*[33]

Sixteenth century:

arrierefais,[34] *arriere-saison,*[35] *avant-coureuse,*[36] *avant-courrier,*[37] *avant-jeu,*[38] *avant-naissance,*[39] *avant-retour,*[40] *avant-venue,*[41] *intérim, intermède, prédiction, préjugé, prélude, préméditer, prénotion, préordonnance,*[42] *pressentir.*

By virtue of the use of these terms, another faculty came to the fore along with the sense of temporal relations. This faculty made it possible more precisely to express the significance assigned to circumstances. These terms served, each in its own way, to distinguish between the main action and its preliminaries, between essentials and incidentals. The flow of events never appears to the human consciousness as a succession every element of which fixes the attention to an equal degree. Events form themselves into groups, into mountains with paths leading towards, up and down them. Those events which are not 'peaks' are at least related to those which are, appearing as paths whereby the summits are approached or left. In Middle French the temporal neutrality or self-contained quality of circumstances disappeared. One process pointed to another, fixing, as it were, its gaze on it. Thus the temporal prefixes had their own share and place in the development of an historical consciousness that aimed at discriminating between the important and the trivial and between preliminaries and what followed.

The vision of space

Henri Bergson pointed out that man robs time of its essential nature to a great extent by projecting the purely inward experience of duration, which has no extension, into space, thereby making of this experience something measurable and homogeneous. This spatialization is not first and foremost the result of a long historical process. It is, rather, at a lower level the consequence of the mastery over time. In order to indicate temporal relations, language makes use of spatial concepts. The idea of movement bridges the semantic gulf between space and time.[43] A study of the view of time prevalent in the later Middle Ages shows how this spatialization of time may be followed as a thread in the history of thought.[44]

In the thirteenth, fourteenth and fifteenth centuries time was viewed to an increasing extent externally, for in that period the awareness of spatial relations was becoming more acute. Among the painters of that epoch a clearer judgment and representation of space had long been in evidence. The Burgundian school of illuminators strove to attain justness of perspective, placing their figures in suitable spatial depth, so that they might stand in appropriate relation to architecture and landscape.[45] Literature likewise showed a concern for clarity in spatial concepts and arrangement. The tale of the raven and the fox reminds us of one of the precisely delineated scenes in the *Book of Hours* of the Duc de Berry:

> Il m'est souvenu de la fable
> Du corbeau qui estoit assis
> Sur une croix de cinq à six
> Toises de hault, lequel tenoit
> Ung fromaige au bec. Là venoit
> Ung renard qui vit ce fromaige.
> Pensa à luy: 'Comment l'aurai-ge?'
> Lors se mist dessoubz le corbeau.   [*Pathelin*, vv. 438–45]

Here the chief centre of attention is the possible execution of the fox's ruse in space. Even the distance of the raven from the ground is stated with scientific exactitude. The fox's 'How shall I get it?' is meant in the physical as well as the psychological sense. The dropping of the cheese is the result, already worked out by the fox, of a mechanical experiment. The wily animal has already calculated the spot on which the cheese must fall.

We have here a tableau rather than a story. Whereas the fable in the manner of La Fontaine creates a psychological atmosphere with its introductory words, here the external, spatial and factual situation is made plain. This was characteristic of the age. The spatial position of people and things was clearly defined. The attention of the poet of the *chanson de geste* was fixed on the inner aspect of action, change and movement, the rhythm and urgency of which communicated themselves to those involved. The sharper eye of the later Middle Ages saw more images and states. The portrait took its place in literature alongside the narrative. The descriptions of Guillaume de Lorris showed a well defined sense of space.

To temporally and spatially undistanced emotional involvement in a train of events was added the faculty of placing a situation in its true perspective. The procession of things and people appeared as a variety of fixed states, pictorially and spatially juxtaposed. The 'belle heaulmiere' of Villon saw the thirty years which stood between her and her youthful beauty not as becoming, flux, transitoriness, experience and progression; instead, she set the image of her youth against that of her more advanced age. Villon was not familiar with the romantic feeling for the passing of time. The poet who sang of the inexorableness of death saw even it as an image. It is not surprising, therefore, that the sense of temporal perspective increased together with this view of the physical world. This manifested itself in the language in a sharper definition of the function and demarcation of tenses,[46] further evidence of a heightened sense of spatial relations. The divisions of time were viewed in juxtaposition, and not as successively lived through.

Space also occupied a predominant place in the idea of the nature of poetry. A poem now appeared as a set of images in the extended physical world. Appreciation of arresting and singular turns of phrase was lacking. The notion of the irreversibility of real time seemed to act as a discouragement to the enjoyment of poetry as an inner experience. The spatial concept of symmetry was applied to verse. Christine de Pisan composed a *Balade rétrograde, qui se dit à droit et à rebours*.[47] The stanzas of an extraordinary ballad by Deschamps are capable of eight different permutations.[48] It may be objected that the entire age should not be judged by such extravagances. Nevertheless, only a turn of mind so acutely conscious of external spatial relations could produce such fantasies.

The new spirit fixed circumstances in space and in spatially conceived time with greater emphasis than was formerly done. They thereby acquired a more definite extension, for two points stand out more sharply from the flow of events; there are the beginning and the end, which serve to delimit a process with regard to others. The beginning and end now gained an intellectual importance equal to that of the circumstance contained by them, because both were necessary elements of the idea of the length of a process. The completeness of a happening apprehended as a whole may be conveyed by indicating its outer limits. This preoccupation with extremities caused the true substance of a process to pale into insignificance, only the two points defining its extension being mentioned. The experience of transition was spatialized in consciousness, and this spatial interpretation of events could loom so large that one no longer spoke of a beginning and an end, but of two ends:

Et Queus li a trestout conté,
De chef en chef la verité...[49]

The notion of a spatial end to process became increasingly frequent and clear. It often happens in human experience that the end of a temporal process and the end of a physically extended object coincide. 'I am at the end of the street' means also 'I am at the end of my walk down the street'. In this way the word *bout* penetrated the sphere of happening, originally foreign to it.

Deschamps:
Se bien servez jusques au bout,
Un coup vendra qui paiera tout.   [II, p. 68]

Cent Nouvelles Nouvelles:
Au bout des diz cinq ans.   [19]

Et n'en fust point venu à bout si, etc.   [67]

Que au bout de quinze jours de dueil il mourut.   [77]

Commynes:

Au bout de trois jours [I, 8 (I, p. 55)]

Avant que fust le bout de l'an. [I, 12 (I, p. 76)]

Au bout d'une heure ouysmes plus grant bruyt à la porte que paravant. [II, 3 (I, p. 113)]

Au bout de troys jours. [IV, 11 (II, p. 76)]

Au bout de deux ou troys jours. [VI, 6 (II, p. 282)]

The duration elapsing between two points in time was turned into distance. If, in the *Condamnacion de Bancquet* by Nicole de la Chesnaye, it was required that the allegorized meals of *disner* and *soupper* should keep a minimum distance of six miles (= six hours) between them,[50] this spatial mode of expression was more than a chance manifestation of the confusion of spatial and temporal concepts, for this interchange was consummated in the direction taken by the general consciousness of time in those centuries. In the above case the unconscious mental attitude inherent in Middle French manifested itself as a conscious play of fancy. Time was now a stretch, seen to be divided into numerous partial stretches. The notion of time spans, of sections of time, gained in prominence. *Espace* was now readily used to denote a definite period; this word was used in the twelfth century, but not until a later age was it employed in its wider sense. Even if the word were not externally recognizable as a loan word, its use would reveal it as such. In many instances it corresponds exactly to the Latin *spatium unius anni*:

Philippe de Novare:
En ce lonc espace de vie.[51]

Deschamps:
Car uns homs puet par longue espace
De temps en femme par la grace
De Dieu avoir enfans planté. [*Miroir de Mariage*, v. 9807]

Par diverse espace de temps. [*Ibid.*, v. 11368]

Cent Nouvelles Nouvelles:
Une grande espace de temps. [57]

L'espace de cinq ans. [19]

L'espace de deux mois. [62]

L'espace de trois ans. [78]

Longue espace de temps. [100]

Commynes:
Dura ce traicté par l'espace de deux moys.   [I, 7 (I, p. 52)]

Ils avoyent desjà faict la guerre par l'espace de cinq ou six mois.
[I, 4 (I, p. 88)]

Et eulx demourèrent sur le marché par l'espace de huyt jours.
[II, 4 (I, p. 119)]

Et fuz l'espace de deux mois allant at venant vers luy pour l'entre-
tenir.   [III, 4 (I, p. 195)]

Lyon, où il s'estoit tenu par l'espace de six moys.
[V, 4 (II, p. 126)]

Et luy a duré ceste bonne fortune et grace de Dieu l'espace de six
vingtz ans.   [V, 9 (II, p. 157)]

By virtue of the frequent use of this word, the language
gained a further means—impersonal, wide-ranging and abstract
in addition—of expressing duration. The 'space of eight days',
now referred to, constituted a frame rather than what filled it;
the span of time mentally overshadowed the eight individual
days, which appeared as though fused into a colourless whole.
There was a tendency to apprehend the span of time intervening
between two events as an empty measure or shell, regarded and
imagined as distance. One learned to see time without its content.
The spatial conception of distance also invaded the realm of time.

Chastellain:
Toutes [les morts] advinrent en un temps et à peu de distance.
[I, p. 341]

Commynes:
Toutes ces promesses se feïrent en moins de troys ans de distance.
[III, 8 (I, p. 222)]

En cest an (MCCCC IIII^{xx} XVIII), que le roy Charles est trespassé, est
finy frère Jheronyme aussi, à quatre ou cinq jours distant l'un de
l'autre.   (VIII, 26 (III, pp. 308–9)]

To the concepts of temporal duration and distance was added
that of the point in time. The latter drew a sharp dividing line
between stretches of time, indicating a state in existence at a
given time, a continuance of things up to a stage at which they
were reviewed. With the word 'point' the course of events was
traced from outside; the arc of the curve they described was
fixed by points. The general significance of 'point' corresponded

to that of 'bout', in as much as it too indicated the end and result of an action of process. It was essentially the climax, the most important moment of an experience. When an event was designated by the word 'point', a short interval of time preceding and following it was passed over, creating the impression that events had come to a standstill.

Cent Nouvelles Nouvelles:
Et sur ce point ilz s'en vindrent disner.   [63]

Il commence de faire ses approuches quand il vit son point ['opportune moment']   [65]

It being the case that, in the later Middle Ages, all terms connected with space served also to designate temporal relations, the word 'between', in addition, was used in a temporal sense.

Monstrelet:
Fut de rechef le service fait par ledit patriarche, mais il y eut une nuit entre les deux services.   [IV, p. 123]

Roi René:
Trois jours le pas durera vrayement,
Au mains depuis une heure aprèz myiour
Jusques à six, et pour grant aysement
Entre chascun des jours ung de sejour.
                          [II, p. 54 (*Le Pas de la bergiere*)]

Des Périers:[52]
Arriva entre les deux messes.   [II, p. 99]

This use of *entre* was unknown to Old French. The notion of 'between' implies juxtaposition and a certain symmetry, which does not exist in time considered as true duration. The phrase *entre deux messes* puts two times on a common plane, in that it draws them together with a single glance. The time here designated by *entre* was formerly expressed by *avant* or *après*: 'avant la messe' or 'après la messe'. One looked either towards the future or towards the past. The idea of 'between the two' embraces them both, corresponding to a clearer idea of space. Although in Old French the word *entre* occurred fairly often in combinations, it never had this temporal significance. In the verbs *s'entr'appeler, s'entr'avertir, s'entrebaiser, s'entrechoquer, s'entrecroiser, s'entreégorger,* etc, it was always a matter of relationship between two persons. This is the specifically Old French use of *entre* in combinations such as occur with uncommon frequency in the work of, for instance, Chrestien de Troyes. In Foerster's *Chrestien-Wörterbuch*[53] we find about forty-five like combinations with *entre* ('one another'.) The objective, spatio-temporal *inter* in composite words (*intercadence, intercéder, interjection, intérim, intermède, interrègne, interroi*, etc) stands in contrast to *entre*, which designated personal relationships and which contained affective overtones. The temporal connotations taken on by the Old French *entre* were, as the examples cited by Tobler–Lommatzsch indicate, an imprecise rendering of simultaneity, as are 'at', 'during', 'on';[54] to this view of things, which did not designate any reference points to 'before' or 'after', the idea of *entrefaites* owed its development, for the concept of 'meanwhile' sufficed for its temporal orientation.

Nor was the ancient language familiar with the idea of interruption. *Intermission* and *intercupacion*[55] appeared in the thirteenth and fifteenth centuries respectively, and *interrompre* in the sixteenth. Before the introduction of these loan words no independent French formations with the same meaning existed. The idea of 'interruption' was expressed in the language of the *chanson de geste* by *sans délai, sans respit, sans repos*. These phrases indicate a mental attitude which identified a process with rhythm and speed. A phrase corresponding to 'an action is interrupted' did not yet exist. The absence of interruption was expressed positively by means of the terms just cited, since the negative concept of intermittence was unknown.

Another concept, the application of which to temporal relations was first made possible by a clearer understanding of the idea of space, was that of 'the remainder'. The word *le reste* did not occur as such until the thirteenth century. The verb already existed in the twelfth century, but we do not find the turn of thought conveyed by *le temps qui reste, le reste du temps, du jour, de l'année*, etc, until the fifteenth century: 'le surplus de sa vie'.[56] Before that no specific and precise phrase for 'the rest of the time' existed. In order to convey the sense of 'the rest of the week (from Friday onwards) plus the ensuing week up to Friday' the author of the *Fabliau des trois Dames et de l'anel* used the following turn:

Iluec fu toute la semaine,
Et l'autre jusqu'au vendredi.[57]

It was possible in Old French, it is true, specifically to express the idea of 'the time remaining'. *Huimais*, for instance, meant 'from today inwards'. But this was an absolute and personal view of time, which represented this remaining time as something future and yet to be experienced. In *le reste du temps* and similar phrases the personal viewpoint of the speaker was implicit. The idea of 'the remaining time' does not refer essentially to future time, but to a stretch of time.

The notion of 'a part' also took its place in temporal thought. The first instances of the application of *la partie* to matters of time are to be found in the fifteenth century.

Monstrelet:
En l'an dessusdit, vint devant le Roy Charles à Chinon, où il se tenoit grand partie du temps, une pucelle josne fille, éagié de vint ans ou environ, nommée Jehenne.   [IV, p. 314]

Roi René:
On pourroit departir le temps par moitié.   [IV, p. 14]

Laquelle en ceste maniere, en bonnes et raisonnables operacions durera la continuacion de sa vie; voire aucune partie de temps.
[IV, p. 74 (*L'Abuzé en Court*)][58]

The official day

Those terms relating to time which the popular tongue created in order to meet the exigencies of an elementary understanding of time and basic orientation stood in no definite intellectual relationship to one another. This lack of connection is evident from their forms. Day and night, yesterday, today and tomorrow, early and late, fast and slow, often and seldom, always and never —all pairs of concepts which to our way of thinking are closely linked—are individual creations from the linguistic standpoint, in as much as the terms 'late' and 'early' are of totally independent provenance. Both were shaped in different moulds; they had originally nothing to do with each other; they stood for two essentially different experiences. Such concepts were gradually drawn together as the awareness of time led to its intellectualization, as the conception of time became more quantitative and as the capacity for continuity of thought increased. Whereas the individual with a primitive sense of time adhered to the natural divisions of time and subordinated his own rhythm of life to them, humanity, in striving to establish the high degree of order necessitated by the advance of civilization, rose above natural divisions of time, felt as experiences, and set up temporal divisions—with some measure of dependence on natural time, it is true, but in such a way that these divisions were properly the work of mankind.

The ordering of time which was introduced to meet the needs of civilization was official time. Its symbol is the clock: it knows neither night nor day, earliness nor lateness, today nor tomorrow. Time became a uniform series of units, points and sections. The spirit of exactitude engendered by this new attitude to time was paid for by the loss of poetry and immediacy in personal experience.

The day provides an example of the influence of civilization, which tended to bring about a conceptual association of natural temporal relations, hitherto subjectively unrelated. The four-teenth-century idea of the day reflected the new spirit. In poetry the rejoicing in the beauties of the daytime evaporated. The naive experience of it gave way to its utilization as a means of reckoning time. In narrative prose the change undergone by the function of the day in the representation of circumstances betrays an externalization of human relations to time.

The modification of the temporal atmosphere, seen by Imbs[59] to be exemplified in the concept of the day, a modification con-summated between *La Queste del Saint Graal* and *La Mort le Roi Artu*, is probably an early episode in the constant process whereby the individual of the later Middle Ages gained distance from the time in which his life was naturally framed. Of the former work he remarks that 'Ce qui compte, c'est la relation vivante de l'homme avec la réalité cosmique qui rhythme le temps et fait naître l'heure',[60] and of the latter: 'L'heure y est ce qu'elle est pour les modernes; un repère, ou une condition favorable à l'action. Dès lors aucune poésie ne peut plus s'y attacher.'[61] The manifestations of time seemed to belie any pleasure in the radiance of the day. It appeared as though the transitory beauty, the birth, radiance and passing away of the day had to give place to its utilization as a temporal unit. It is hard to believe that in this age, which according to Huizinga was so receptive to the concrete beauty of life in all its colourful variety, the alternation of day and night was not an impressive experience also. But poets were not inspired by the course of the day and its radiance. Possibly it was accepted so axiomatically as a natural phenomenon that it was not mentioned. It may be that its connotation had shifted from that obtaining earlier. The day was the light emanating periodically from the sun. Those things which went to make the day were objects and processes illuminated and warmed by the sun. The more distinctly they were seen and named individually, the less that phenomenon to which they owed their visible existence was mentioned, a phenomenon which was present in all and yet in none. Thus the day appears to us in the texts of the time as a numerical entity rather than as an experience. It seemed to be something faceless. The language was unaffected by the natural experience of the alternation of day and night. Both appeared with increasing prominence as sections and boundaries of time, whose content and particular colour as an experience played a secondary part. Just as day and night were equated and ranged together in thought, so were they ranged together in the spoken tongue. In official language they were reduced to a common level. Even in Old French we find the phrases 'de jour et de nuit' or 'par jour et par nuit' in translations; they were doubtless formed on the model of the Latin *de die et nocte* and *per diem et noctem*.

Psalter:[62]

Par jurn et par nuit.	[I, 2]
Par jurn et... par nuit.	[XXI, 2]
Par jurn et par nuit.	[XXXI, 4]
Par jurn et par nuit.	[XLI, 3]
Par jur... par nuit.	[CXX, 6]

Quatre Livres des Rois:

Par nuit é par jur.	[Book I, xxv, 16]

Et ne suffrid pas que les oisels entamassent les cors de jurs ne les bestes de nuiz.   [Book II, xxi, 10]

Archives Wallonnes:[63]

Par jour ou par nuit.	[13]
Soit par nuit soit par jour.	[36]
Ne par jour ne par nuit.	[137]
Par jor ne par nuit.	[137]

Par nuit et par jour.	[164]
Par jour et par nuit.	[164]
Par jor... par nuit.	[194]
De jour ou de nuit.	[13]
De jour... de nuyt.	[14]

In a later age these phrases were of very frequent occurrence in general literature. In the vast majority of instances the word *jour* preceded *nuit*—rarely the reverse. This is in accord with Spitzer's remarks on the order 'night and day'.[64] This order is illogical and affective, for the word with the stronger appeal to the imagination is placed first: ' "Night" was intimately associated with "evil"; in the popular mind, "night" was a source of danger, something "dark"; one did not know whether one would escape from the darkness of the night or whether one would ever awaken again.' No doubt other considerations played a part. The daytime appeared as something familiar and 'right'; in short, as a state of normality. The actual light of the sun was thought of in the same way. The Beyond as imagined as day and not as night, presumably because, in popular thought, the day was not a natural phenomenon comparable to that of night but was regarded as something normal and taken for granted. The night, on the other hand, appeared as something exceptional which sprang to mind more readily than the usual. Evil made a stronger impression than good, war was more vividly imagined than peace; the same held good for death and life, night and day. In Old French the form *nuit et jor* predominated; by almost always having *nuit* as the first word, it stood in sharp contrast to the form: *de (par) jor et de (par) nuit*. We quote some examples of the older form, which lay no claim to carrying conviction by weight of numbers:

Old French romances:

Et ou est ores li vales
Qui neut et jour m'aloit dixant.[65]

Dites qu'estes donee au dieu mestier
En tel labor et nuit et jor
Por dieu prier.[66]
Por vostre amor
Li cuers mi sautele
Et nuit et jor.[67]

Dame, vostre rentier
Me faites nuit et jor.[68]

Roman de la Rose:
L'iaue est tousdis fresche et novele,
Qui nuit et jor sourt à grans ondes. [v. 1539]

              tu accompliras
Nuit et jour les commendemens
Que je commande as fins amans. [v. 2050]

Que nuit et jor sans repentence
En bien amer soit ton penser. [v. 2244]

Les ténèbres où li cuers gist
Qui nuit et jor d'amors languist. [v. 2760]

Car Vénus l'avoit envaie,
Qui nuit et jor li emble
Boutons et roses tout ensemble. [v. 2863]

In contrast to these examples, where *nuit et jour* approximated to the sense of 'always, constantly', *de jour et de nuit* had the more factual sense of 'by day and by night'. It is less impressionistic and rhythmical, but more literal in its temporal indication. Time was not an impression but a fact, and was represented as such in language. Its purpose was less to indicate a mental attitude than to point to a circumstance independent of one's personal opinion.

The essential nature of official time is that it stands between individuals, both binding and dividing them. It is the medium of mutual understanding and communication. The existence of official time became increasingly real in the fourteenth and fifteenth centuries. What men had to say to one another had to be filtered through this medium, which imposed distance on human relations and restraint on affectivities. To make this clearer, let the reader think back to the *rapport*, as represented in the *Chanson de Roland*, between Charlemagne and his nephew. There, no objective and officially binding time stood between the personages, whose relationship to one another was depicted as something altogether inward. They appeared as though free of any mutually binding external obligations. What was requested and granted was the natural expression of the sentiments shared by all, which alone carried weight. Ethos and affectivity were so predominant that there was no room for any possible allusion to the strictly lawful relationship existing between the parties. This inward quality of personal relations generally obtaining in the *Chanson de Roland* and the national epic was somewhat modified in consequence of the more concrete circumstance of the death of Garin de Lorraine. Here the expiation of the murder appeared as a judicial matter. In the epic one of the most important considerations was the interplay of personal relations; those concerned were clearly aware of their own rights. Their attention was engaged by the points of contact between their own rights and influence and those of others. They attached much importance to honouring of temporal commitments. Thus the official day, mutually agreed on by the parties, played an important part in *Garin*:

Je pris le jor, et si fu bien assis.	[v. 146]
Li Loherens au jor les atendi.	[v. 164]
Prenez un jor, s'il vos vient a plaisir.	[v. 264]
Li jors failli, un autre en avons pris.	[v. 301]

The notion of the official day is met with elsewhere in ancient French literature, but the idea of time, thought of as standing between men, was not incorporated into thought until a later age. Evidence of the blatant and automatic manner in which the citizen of the later Middle Ages viewed time as a mere counter is provided by the extremely numerous temporal agreements entered into by the personages of the *Cent Nouvelles Nouvelles*, in which the proposal and acceptance of a point in time had a purely external significance:

Ce desiré jour fut assigné.	[1]
Il part de léans, et prend jour à demain de retourner.	[2]
Jour assigné. Au jour nommé.	[3]
Qu'elle lui baillast et assignast jour. Bailla journée. Bailler jour. J'assignay le mauldit jour.	[4]
Elle luy baille jour. Il accepte la journée. Il ne fauldra pas a sa journée. Il a jour assigné.	[9]
Elle luy baille jour, à douze heures de nuyt.	[15]

This example shows how the word *jour*, when used thus, completely lost its proper meaning.

Elle mesme avoit l'heure et le jour assigné.	[28]
Leur baille jour et heure.	[34]
Et fut pris jour de paier a deux termes.	[43]
A jour determiné.	[89]
Ung jour qu'elle nomma.	[93]
Nous prendrons et assignerons un jour.	[98]

Treated in this way, the day became simply a means of indicating a time, or was turned into a short stretch of time or even a point. It had no validity as a process *per se*. The day was thought of only from the standpoint of its utility as a means to the orientation and ordering of life. It acquired an objective value which was the same for all. The awareness of the temporal relationship in which the day stood with regard to other periods became more acute. It was now a matter of selecting and earmarking certain days from among the countless others that succeeded each other through the year or years, according to their importance or by virtue of some circumstance with which they were associated, a necessity which can be gauged by the increasing frequency with which the phrase *le jour de la passion* occurred as the period of Middle French approached.

This phrase entered French from biblical language. In classical Latin the genitive of an abstract word, as in *dies irae, ultionis, tribulationis*, was unknown.[69] Examples of this use of the genitive in Old French, direct imitations of the Latin model, are to be found solely in the ecclesiastical domain:

Guillelme:
Al jur de grant juis.	[v. 1426]

Aliscans:
Au saint jor del noal.	[v. 548]
Au jor d'ascension.	[v. 7112]

Psalter:
El jurn de tribulation.	[XIX, 1]
Les jurz des nes.	[XXVI, 19]
Le jurn de temptaciun.	[XCIV, 8]

On the other hand, independent formations which named a day after an ecclesiastical event, appeared sparsely and hesitantly. In the *Archives Wallonnes* of the twelfth and thirteenth centuries[63] we find: 'Le jour del Bouhourdic',[70] 'au jor de me mort'.[71] *Aujourd'hui*, which first occurred in the thirteenth century,[72] may probably be included amongst these genitives. Links between the two terms 'day' and 'today' are to be found in the very earliest examples of French literature:

Passio Christi:
Oidi.   [v. 292]

Oi en cest di.   [v. 299]

Alexis:
Oi cest jurn.   [v. 542]

Roland:
Hoi cest jur.   [vv. 2107, 2751, 3100]

These forms alternated with *hui*, yet we cannot say for certain why sometimes one form is to be found and sometimes the other. No doubt *oi cest jor* was more emphatic than *oi* alone: 'Current actuality . . . is distinguished from ever-present possibility by means of the insistent demonstrative.'[73] The immediate predecessor of *aujourd'hui* was *cest jur de ui*, found in the Old French translation of *Les Quatre Livres des Rois:*

A cest jur de ui nen iert nuls ocis.   [II, xi, 13]
Sil frai a cest jur de ui.   [III, i, 30]

It was a form that did not last, possibly because the form *aujourd'hui* was more euphonious. The latter first occurred in early thirteenth-century archives; it is hardly to be found elsewhere during that century:

1219:
Totes autres quereles qui ont estee ou puissent avor estee entre iaus de cateil u de meules juskes au jor de hui.   [Le Proux,[74] No. 6]

1226:
De totes altres coses que duskes a cest jor dui ont este entre li e monsignor Willaume.   [Tailliar, No. 27]

1241:
Et cilh anz commence a ior dui cest a dire a ior de mon sanhor sen mark Lenwangeliste.   [Wilmotte,[75] *Romania*, XVII]

1255:

Toutes les trives ki ont este prises par eskevins dusques au jour dhui.   [Tailliar, No. 138]

1258:

Les verites ke nous en avons oies dusques au jour de huy.

[Tailliar, No. 146]

1264:

Li eschevin dient dun meisme acort por lo miols quil sevent au jor de hui.   [Tailliar, No. 176]

1265:

Tous les proufits des linaiges des hales de le vile de Douay, ki aujourd'hui sunt.   [Tailliar, No. 117]

1265:

Ains doivent estre a ceus qui les i ont euwes juskes aujourd'hui.

[Tailliar, No. 177]

1283:

Que tout chil qui se sont mis en prison jusques au jour dui, sil ne se sont racatet dou jour de hui en demi an.   [Tailliar, No. 227]

1286:

Et que tout li yretage qui aujourdhui sont en wages, soient racatet doujourdhui en II ans et II jours.   [Tailliar, No. 231]

1286:

Des yretages qui ont estet en wages jusques aujourdhui.

[Tailliar, No. 231]

1289:

Toutes les debtes qui prises sont et ont estet jusques aujourdhuy.   [Tailliar, No. 237]

1294:

Pour loccoison des acques devant dis, que il ont ou leurs predecesseurs ont fait jusques au jourdui.   [Tailliar, No. 243]

Except in legal texts, the expression is hardly to be found in the literature of the thirteenth century. Littré cites an example from the *Roman de la Rose*:

Mal vit ajorner le jor d'ui.   [v. 11619]

The fact that *aujourd'hui* occurred with such frequency in records at a time when it was scarcely used in other literature, as also the manner in which it was employed, clearly indicates the nature of this word. It was official, legal and factual. It had nothing to do with time as an experience, but its purpose was to fix a particular point on the scale of temporal units. Spitzer realized this when he considered the development of the word to be the expression of a new temporal perspective.[76] *Aujourd'hui* represented a point of repose in time and a temporal viewpoint. The examples we have quoted demonstrate how this term established a vantage point in time, whence all previous occurrences and all future possibilities were surveyed within certain temporal and circumstantial limits. It was first used, not with a glance at, nor in contrast to, 'yesterday' or 'tomorrow', but with reference to a much wider temporal field of view.[77] The person using the word saw time as a spatial succession and series, in which the present held an especial place as a point of view. This use was not a popularly current one. *Aujourd'hui*, therefore, has not become a strictly integral part of French dialects up to the present day. By means of this term one takes stock of what is in some sort present up to a point in time. This placing of the day in a wider context, which might embrace years and decades, explains the use of the word in the sense of 'nowadays', a sense that was much more frequent in Middle French than the strict one of 'today' when used in antithesis to 'tomorrow' or 'yesterday'. In the work of Eustache Deschamps, that *laudator temporis acti*, dozens of instances may readily be found in which *aujourd'hui* means 'the present'. Thus in the Middle Ages the term moved increasingly on to a higher plane than that of immediate everyday necessity. The 'today' was, on the one hand, the day which could easily be distinguished from others by means of its date and, on the other, the present which claimed the attention of a sharpened historical sense.

Time as a condition

In historical and other literature the turn of mind, predominant in former centuries, whereby events were viewed as the natural consequence of the attitude of those concerned, began to lose ground. Roland had to die, because his death had its roots in his character. This was an inner necessity. This notion of an inner necessity was now lost sight of. Fate was now increasingly regarded as the joint working of external circumstances. A story reported by Chastellain[78] shows how, in contrast to the conception prevailing in the *Chanson de Roland*, death could become purely a question of time.

A young man who had killed another was in the prison of Charles the Bold in Bruges. The murderer's associates succeeded in placating the victim's relatives, who had sworn revenge. Charles, however, did not liberate the murderer. So that the case might serve as a warning example, he ordered a court official to have the miscreant hanged at eleven o'clock on the morrow. The official then informed the murderer's kinsmen, who were trying all means to secure a repeal of the sentence, of the circumstance. The execution was postponed by the official, on his own authority, until three o'clock in the afternoon, 'par espoir qu'en ce pendant le duc pourroit remitiger sa rigueur'.[79] It was now a matter of seeking out and mollifying the Duke, who was absent from his residence. A repeal of the verdict was secured—after the sentence had already been carried out.

Everything in this story turned on the all-important question of time. The passing of time and the approach of the hour of the execution made themselves painfully felt: 'Le temps coula et monta tousjours à haute nonne, et trouva sa peine perdue et sa diligence faite en vain.' This awareness of time as a force determining events introduced an important element of external tension into contemporary narrative. Time had a great influence on the outcome of events. It was the deciding factor in the most weighty matters. Chastellain also represented the decision in the quarrel between Warwick and Henry in England as being dependent on time.[80] Henry, an associate of Charles of Burgundy, was required to beat Warwick in combat; he duly confronted him, but not until 'trois jours après la bonne heure'. The narrator then speculates on what would have happened had the appointed time been adhered to.

Time, however, was acknowledged as a force to be reckoned with even in the most trivial aspects and happenings of daily life. The Frenchman of the later Middle Ages stood with both feet firm on sublunary reality, the laws and manifestations of which became more familiar to him in proportion as the image of eternity faded. The image of the world was clear and objective: one knew the function one had to fulfil in it, and one knew and called by name those indispensable concomitants of successful action, space and time. Much depended on the favourable interplay of both. They were indispensable in a twofold sense. On the one hand, they were the necessary media of process in general and, on the other, precisely appropriate temporal and spatial conditions were a prerequisite for meaningful or profitable action. There was now an acute awareness of these and of the dependence of events on them. The literature of the period of Middle French reflected this awareness of the functional role of space and time. In the *Cent Nouvelles Nouvelles* everything hinged on spatial and temporal possibilities and their utilization. While the psychological element remained uncomplicated and obvious, it being the case that the characters were emotionally primitive and underwent no inner changes, the narrator's attention was fixed on the external circumstances, the where and when. This 'where and when' occupied a place of prime importance, assuming as it did various guises—absence, lack of time, a favourable opportunity, a sudden return, concealment, the keeping or otherwise of an appointment or the making of one. Successful action appeared as the result of the favourable co-operation of space and time, as the coincidence of the right place with the right moment. The now acute awareness of the external media of events seems to have found its expression in language in the phrase *en temps et en lieu*, frequent in Middle French:

Chrestien de Troyes:
Si lor comance une reison,
Qui vint an lieu et an saison.   [*Cligès*, v. 2277]

Quant il ont eise et leu et tans.   [*Ibid.*, v. 3864]

Roman de la Rose:
S'il en éust et tens et leu.   [v. 13473]

Si tens et leu avoir péust.   [v. 13466]

Dis dou vrai aniel:
Car sens gouvrenés de raison
A en tous poins lieu et saison.   [v. 22]

Deschamps:
En temps et en lieu.   [*Mir. de Mar.*, v. 6302]

En lieu, en temps et en saison.   [*Ibid.*, v. 9365]

Chastellain:
Pour satisfaire aux autres matières opportunes en leur lieu et temps. [II, p. 115]

Saintré:
Promesses... lesquelles en temps et en lieu se doibvent accomplir. [Chap. 15]

Cent Nouvelles Nouvelles:
Et, pour y bailler remède convenable, ne restoit mais que temps et lieu. [1]

Trouver temps et lieu. [4]

Si j'eusse temps et lieu. [18]

En temps et lieu. [86]

Comment elle avoit, si à elle ne tenoit, si bonne habitude et opportunité de temps et de lieu. [100]

Le lieu, le temps, toute opportunité nous favorisent. [100]

Monstrelet:
Car, si onques il fut lieu et temps de prescher la justificacion et loyaulté de monditseigneur de Bourgogne, il en est ores temps et lieu. [I, p. 184]

Ausquelz le capitaine respondi qu'il les excuseroit en temps et en lieu. [III, p. 178]

The idea of time had now become objectivized to such a degree that 'having time' was seen as a condition for doing or not doing something. The phrase 'to have time', which in developed languages exists as a matter of course, was alien to the primitive view of time. Time, if one is to 'have' it, must be something external. The phrase implies that of tenone does not have the time one needs. Apart from the will and the ability to perform an action, one requires time in which to do it. To the medieval mind one's personal attitude as a factor in action was not in itself sufficient: such action was subject also to external circumstances. Having time at one's disposal was a matter of uncertainty. The following examples show how little this 'having time' was represented as a fact by the writers of the period of Middle French.

Deschamps:
Compter me fault, se temps ay et espace ['time'] [I, p. 79]

Chastellain:

Pour seule ardeur de vengeance dont il prioit Dieu pour avoir temps seulement de la mettre à fait.   [I, p. 51]

Et convenoit avoir grand temps, avant que tout fust mis en ordre.
                                                            [V, p. 420]

Monstrelet:

Mes lectres portans date de l'XIe jour de juing derrenièrement passé, lesquelles vous povez bien avoir veues en digne et souffisant temps.   [I, p. 16]

Commynes:

Par la mauvaistié de ung ou de deux, ne se doit laisser à faire plaisir à plusieurs, quant on en a le temps et opportunité.
                                                      [II, 3 (I, p. 116)]

Il taschoit à tant de choses grandes, qu'il n'avoit point le temps à vivre pour les mectre à fin.   [III, 3 (I, p. 189)]

Ceulx qui ont eu espace et temps de vivre.   [V, 18 (II, p. 214)]

Il ne se fault point tant haster et a-l'on assez temps.
                                                      [V, 19 (II, p. 217)]

Saintré:

Et autres sans nombre, que je bien nommeroye se j'avoye temps.
                                                             [Ch. III]

After the use of abstract forms of expression had been stylistically liberated by the allegorizing tendency of the later Middle Ages—that is to say, when greater freedom in combining terms had become possible—one was able to say, instead of *avoir temps,* such phrases as *le temps le permet,* in which time appeared as a personal power on whose will the happening of the event depended:

Alain Chartier:

Or je te pry, tant que nous sommes sur ce pas et que le temps le peut souffrir, que tu me vueilles declairer... quelz gens tu reputes dignes de deffendre le bien publicque.[81]

Chastellain:

Là où le temps me donnera occasion de parler plus amplement.
                                                            [I, p. 217]

The idea of 'having time' was alien to Old French. A mental concept closely approaching it was *avoir loisir.* This, however, did not mean 'to have time' in the narrow sense, but had the general significance of 'to have the opportunity, permission, facility, to be allowed to act, give, etc, without constraint'; it thus contained affective overtones. In the ancient tongue, time, place and authority were designated synthetically by one word, whereas today we differentiate them analytically as separate conditions. In that age, what was designated by the word *loisir* was not connected with the idea of 'time'. The phrase *avoir temps*, with its more restricted and precise meaning, did not exist, because the conscious awareness of time as a condition governing a process did not exist. *Loisir* was not simply another word for 'time' but something quite distinct from it. In the *Chanson de Roland* it did not designate a span of time but a rhythm, the rhythm with which time was informed; it was thus an affective as much as a temporal term:

> *Roland:*
> Li Emperere en tint sun chief enclin;
> De sa parole ne fut mie hastifs,
> Sa custume est qu'il parolet à leisir. [v. 141]

The expression *parler à loisir* meant 'to talk calmly, without disturbance or interruption'. *Loisir*, therefore, as is evident from its use in ancient French texts, was never as strongly represented as a condition as was time in *avoir temps*.

The diversity of times

'Les temps et les hommes sont divers,' Chastellain remarked. It is true that, even in such an early work as *La Vie de Saint Alexis*, the past and the future were set against each other, but the idea of a succession of ages that were different and distinct from one another was unknown prior to the thirteenth century. Before then there were no epochs, no awareness of the diversity of temporal phenomena; nor were these phenomena appreciated as being peculiar to a given era. The historians of the late Middle Ages had a clearer image of the past and were aware of the context in which historical situations were set. The absolute view of time paved the way for an enhanced capacity for the evaluation of the relative nature of historical developments. Truth was now not only perceived as something unique and timeless, but the manner in which it was refracted as it passed through the prism of the ages was also noticed. Those things that were separated by time seemed, in addition, unlike in their outward and inward aspect. Time was now a force that was held to be the cause of the peculiar qualities of men and circumstances in a particular age. Historical facts were recounted with such observations as 'comme le temps portoit ainsi', 'le temps l'a ainsy apporté,' 'ainsi le rapporta le temps'. Phrases such as *selon le temps* enjoyed a great vogue. Events were now judged not only by an invariable criterion; temporal circumstances were also taken into account. Human perception and judgment had gained greater mobility; action had become more flexible and adapted itself to the course of events. In the evaluation of things, customs, circumstances and actions the temporal element played an important part.

Froissart:
Selonc le temps se couvient ordener
Et mettre en li raison, sens et mesure,
Car on poet trop perdre par soi haster.[82]

Monstrelet:
Les hommes... treuvent ou ymaginent, selon la qualité du temps, aucunes nouvelles manières qui leur semblent nécessaires.
[IV, p. 126]

Chastellain:
Selon le temps et les diverses affaires qui vous peuvent survenir.
[I, p. 44]

Cestui, non veuillant fuir l'entendement de son père par moins, ains plustost l'approcher par plus vive signification et plus ague, selon le temps que veoit, prist et mist sus pour enseigne perpétuel de sa maison le fusil. [II, p. 8]

Autres aussi qui plus veoient parfont et considéroient le temps qui régnoit. [II, p. 81]

Considérant la variation des divers temps qui rendent diverses estrangetés. [V, p. 198]

Vu le temps quel il estoit et quel il avoit este depuis le couronnement de ce roy Loys. [V, p. 329]

De quoy le roy véritablement, par duresse et par quérir en luy ce que ne séoit pas bien, selon le temps d'alors, estoit seule cause.
[V, p. 419]

Les temps, entends bien, veullent estre congnus et appètent que les cœurs des hommes s'arrèglent à leur rendition, soit par patience en choses dures et difficiles à muer, ou par subtile dissimulation jusques au terme que les temps mesmes font muer les choses.

[VII, p. 130]

Roi René:
Pour me gouverner selon le Temps
Que pour lors avecques moy avoye.   [IV, p. 137]

Commynes:
Lui souldoya secrettement quatorze navires de Oustrelins bien arméz, qui promectoient le servir jusques ad ce qu'il fust passé en Angleterre et quinze jours après. Ce secours fut très grand selon le temps.   [III 6 (I, p. 212)]

Jehan de Paris:
Les cent barons et les cent pages, en belle ordonnance, vindrent devers le roy à Vincennes, habillez si honnestement que estoit merveilles et belle chose à veoir, selon le temps qui pour lors couroit.[83]

Judgment passed on human action now appeared to be conditioned, no longer in conformity with a rigid moral principle, but in accordance with temporal circumstances. Commynes reported of some soldiers that, at night, in consequence of a rumour, they mistook a cluster of thistles for enemy spears. At daybreak they realized their mistake:

Et en furent honteux ceulx qui avoyent dit ces nouvelles, mais le temps les excusa, avec ce que le paige avoit dit la nuyct.

[*Mém.*, I, 11 (I, p. 74)]

This stylistic raising of time to the status of an agent is in this instance not a poetic or didactic allegory, but a new means to the linguistic representation of a logical relationship.[84]

It was thus often the case that a circumstance could not be properly understood unless it was seen in the light of its temporal context. One of the most significant aspects of fifteenth-century thought was the ability to adapt to a circumstance in this temporal context, whether it was a matter of historical and theoretical adaptation in judgment of things past, or of practical adaptation in the struggle of life. Contemporary literature was studded with phrases indicating the demands of time, such as 'à l'exigent du temps',[85] or 'à l'exigent de l'heure'.[86] In a letter of Pope Gregory

XII addressed to the king of France and the University of Paris, and translated by Monstrelet, the exigencies of time were set against the immovable standpoint of the law:

> Car au droit il ne loist pas toujours soy incliner, mais on doit toujours avoir raison du temps et de l'utilité.   [I, p. 148]

The necessity of having regard to temporal circumstances and of adapting oneself to certain changes in one's environment was conveyed by the formula 'to move with the times':

> Le temps s'en va et on demeure;
> Sy dict on souvent qu'à toute heure
> y fault aler avec le temps.[87]

This formula, however, did not contain the present-day connotation of progress.

Against this proponent of opportunistic worldly sagacity, Nouvelle Prudence, who moved with the times and was able to adapt herself to them, Alain Chartier in his *Dialogus familiaris* set the (so to speak) timeless wise man, who defended enduring and everbinding values. This abrupt and acute confrontation of attitudes lays bare to us the spirit of the times in its characteristic manifestations.

> *L'amy* Les punicions passees sont de leger mises en obly, et n'y a aussi comme nulle cautelle pour pourveoir au cas a venir. Aincoys aujourd'uy vit chascun avecques le temps tel qu'il avient, et tant comme l'ung de nous peut succeder sans periller, nous cuidons avoirs vaincu fortune de tous points.
>
> *Le sodal* Et penses tu que ce soit de saiges gens?
>
> *L'amy* Mais plus fort on dit que les sages gens de maintenant passent ainsi de temps et vivent avecques le temps.
>
> *Le sodal* Or se donnent bien garde que le temps ne les passe. Car il ne va pas droitte voye qui ne regarde son chemin d'avant, par ou il doit passer. Autrement il se pourroit de leger blessier, ou tresbucher en quelque abisme; car c'est veritable prudence que de bien considerer et penser les fins des choses.
>
> *L'amy* Quant autre temps advient les gens treuvent autre maniere de vivre, et varient leurs meurs et leurs loix par my le monde, et ainsi vivent autrement et autrement selon les temps.
>
> *Le sodal* Les temps passent sans sejourner, et les fortunes des lieux se muent, et si se alterent les hommes. Mais icelle prudence, qui est droicte chartiere et conduiseresse et moyenneresse des vertuz, demeure tousjours incommutable, eternelle et non mesurable...[88]

This changed contemplation of things in time was a source of distress if it seemed that the order of things frustrated one's efforts or foresight, or if it did not appear as a logical development, but as the working of chance, of an inscrutable fate, or of a sudden whim of mutability: 'Here today and gone tomorrow.' This face of time was now perceived.

Huy hoste, demain fait bailler.

Tel est par luy en hault monté,
Qui landemain est debouté,
Ceulx qui l'ung jour ses amis sont,
Ont par luy landemain bont.

Or soustient l'un, tantost luy fault.
Huy boute l'un, demain le tire.   [Roi René, IV, pp. 100f.]

Time showed an infinite variety of faces. It embraced all and united within itself the most extreme contradictions: truth and falsehood, good and evil, greatness and smallness, avarice and generosity, indolence and industry, envy and love, licence and chastity, and so on. It contained all possibilities, whether for good or ill. This many faceted time was allegorically represented by René d'Anjou in his *L'Abuzé en Court*. Time at court—that is, the rapid swings of fate and even of the nature of things in general—was an image of the outside world; time, like the world, betrayed those who depended on it.

In order to express the changing forms of time, greater use was made of those linguistic resources neglected in Old French. The combination of a temporal term plus the genitive of a thing was frequent.

Chastellain:
A l'heure de besoing   [II, p. 141]

En temps de paix   [II, p. 143]

En temps de guerre   [III, p. 8]

En temps de besoin estroit   [III, p. 442]

En temps de murmures et de rumeurs   [V, p. 422]

Commynes:
En temps d'adversité   [V, 7 (I, p. 143)]

Durant le temps de prosperité   [V, 11 (I, p. 166)]

En temps de paix comme en temps de guerre [VI, 12 (II, p. 335)]

In *La Vie de Saint Alexis* we find:

al tens ancienor, all tens Noe, al tens Abraham.

In Old French this form was used almost exclusively with proper names. The time was named after a person. This was a general, nebulous and meaningless indication of time. This new use of the genitive, however, served to denote the factual characteristics of an age. The latter was designated by its content, its significance and its individual nature. This also was an objectivization of the concept of time. One was now aware of the temporal limits of things and values, of their ephemeral validity. What was now insignificant did, nevertheless, have its importance in a former age. Those things which no longer existed were once realities for the men of another epoch. This awareness of the varying characteristics of different ages developed. The need to express the temporal limitations of a thing seems to have had its origin in a linguistic idiosyncrasy of those centuries. This consisted in a tendency to use the preposition *pour* in indications of duration and points in time.

Chastellain:
soy mettre aux champs, qui estoit la chose que pour celle heure ils désiroient devant toute autre.   [II, p. 13]

Commynes:
Et me semble que pour lors ses terres se povoient myeulx dire terres de promission que nulles autres seigneuries qui fussent sur la terre.   [I, 2 (I, p. 13)]

Telz ay-je veu le roy, ledict conte de Charollois, pour le temps de lors, et le roy d'Angleterre et autres plusieurs.   [I, 12 (I, p. 79)]

Pour ce temps les Oustrelins estoient ennemys des Anglois et aussi des François.   [III, 5 (I, p. 203)]

Pour lors n'estoient point estiméz comme ilz sont pour ceste heure.
                                                       [V, 1 (II, p. 105)]

Pours le temps que je l'ay congneü, point n'estoit cruel.
                                                       [V, 9 (II, p. 154)]

Pour le temps que j'estoye à luy.   [VI, 1 (II, p. 242)]

Jamais homme ne se trouva pour luy, au moins pour le temps de lors dont je parle.   [VI, 3 (II, pp. 260–1)]

Cent Nouvelles Nouvelles:
Une fueille d'oseille qui pour l'heure de adonc estoit couverte et
soubz la neige tappie. [19]

Quel que mal content qu'il fust pour ung temps. [24]

Or voy-je bien qu'il fault que je vous abandonne pour ung espace.
[26]

This *pour* served to express the idea of temporariness. Whereas
à simply dated a circumstance, *pour* represented something as
appertaining to a time. It was readily used on account of its
relevance to cultural and historical vision. It was employed to
indicate how things were 'then', and was thus a symbol of the
overcoming of the medieval anachronism. It may be said that a
cultural and historical sense now indeed existed. Landmarks in
history were recognized, and things were seen in relation to them.
The facts of history were now seen less as standing alone and
unconnected, but rather as related to one another by the age to
which they belonged. In the minds of those who took stock of the
times the idea of a state was added to that of a succession. History
was viewed also as a stable circumstance. It is not surprising that
words having the meaning of 'circumstance' were frequently
employed in Middle French. These were *estat, cas, point, temps,
circonstance*. The age in which one still lived was seen with
particular clarity. Both mentally and in the language, it stood
out with increasing plasticity against other ages. The historian
drew an ever sharper line of demarcation between the present
and the past. Now, for the first time, there was truly a present
in the cultural, moral and psychological sense. The language
offered rich resources with which to indicate this present.

Deschamps:
Le temps qui est. [II, p. 40]

Ce temps. [I, p. 130]

Presentement. [I, p. 131]

Le temps qui court. [I, p. 179]

Or. [I, p. 205]

A present. [I, p. 207]

Ce monde present. [I, p. 246]

L'aage present.   [II, p. 43]

Le temps present.   [II, p. 6]

Le temps que voy.   [II, p. 44]

Pathelin:
Le temps qu'on voit presentement.   [v. 132]

The word *temps* in the sense of 'the present' was used without further qualification; this usage is still current today ('the needs of the time', 'the duties of the time', 'the voices of the times', etc).

Monstrelet:
La qualité et malice du temps.   [I, p. 147]

In the consciousness of the later Middle Ages not only were the present and the past distinguished by their different historical aspect, but, in addition, there existed a painful awareness of the gulf between what was and what was no more. The exploits of the heroes of the *chanson de geste* were never felt as something past, they lived on into the present; they formed a part of it and had that tragic quality of what is irrevocably lost. One now became aware of this tragic quality for the first time. The past appeared as a vanished world. Villon realized the existence of this gulf between the present and days of yore, both in his own life and—with wider-ranging vision—in the course of the world. The present acquired a quality of greater reality and certainty, and the past became problematic. It could no longer be said to exist.

Time as a remedy and aid

The lofty unconcern of the medieval mentality for the exigencies of the moment—understandable when it is realized that its attention was held by that which was unaffected by time—now disappeared. Contemplation of the inner, unalterable nature of things gave way to a preoccupation with the workings of chance and other external phenomena. Differences among men and circumstances were seen as a development. A circumstance which, when viewed in the light of absolute and timeless criteria, appeared simply as good or evil, beautiful or ugly, great or small, seemed, when the temporal factor was taken into account, to be unripe, imperfect, fully developed, capable of development, or mature: in short, as a stage in some progression. That which did not yet exist might, with time, come into being. Time thus rendered assistance to man—what was denied by the present, that the future might grant. The attainment of an end became a matter of time. The awareness of the silent and unceasing workings of time was reflected in numerous new turns of phrase. The person with an eye to his own interests knew how to use all means to the attainment of his goal. The changes of circumstance brought about by time appeared to him to further his purposes.

Chastellain:
Moyen toutesfois y fut trouvé depuis, par longuesse de temps.
[V, p. 203]

Le roy Loys pour ce temps-cy se tenoit encore en Touraine, là où tousjours avecques le temps et de jour en jour avisoit soudainement de maintes estrangetés. [IV, p. 195]

Cent Nouvelles Nouvelles:
Gerard et Katherine par succession de temps s'entramèrent tant fort et si loyalement qu'ilz n'avoient qu'un seul cueur et un mesme vouloir. [26]

Practical wisdom demands that future contingencies arising from present action be taken into account:

Car il ne leur challoit de prime face de veoir leur seigneur affoibly d'une telle ville ne leur sens ne congnoissance ne alloit pas assez avant pour congnoistre le prejudice qui leur en povoit advenir à traict de temps. [Commynes, *Mém.*, V, 17 (II, p. 200)]

A longueur de temps [le foible]
aura sa raison, si la court… n'est contre luy.
[*Ibid.*, V, 18 (II, p. 213)]

There now existed an ability—one which might be described as scientific by virtue of its contemplative, if not its methodical, quality—to discern the gradual changing of things, and to follow a process of becoming through its various stages. The slipping by of the hours and days, which went on slowly and by degrees, and which was perceived by a sharpened awareness of small changes, likewise found its expression in language. Men, things, circumstances and institutions were never so much objects of perception as now. Formerly they were seen as abstract and symbolic; there was a greater awareness of their qualities of inwardness, immanence and changelessness than of the manner in which they actually manifested themselves at different times. Now the thing itself was contemplated rather than the idea of it. As a result of this more penetrating observation, everything seemed in a state of flux.

Chastellain:
Tousjours depuis de degré en degré, par succession de temps, il commenca à florir et à prospérer en ses faits.   [II, p. 180]

Le temps couloit tousjours avant et se multiplioient les affaires en ce pays de Hollande.   [III, p. 81]

Comme le temps va tousjours coulant, et que bons usages et bonnes mœurs cessent et se changent, et muent en mauvaises et se refroidissent par longuesse de tems.   [V, p. 254]

The inmates of the besieged fortress, awaiting assistance, became painfully aware of the passage of time:

Et tousjours s'avançoyt le temps, en la detresse de ceulx de Novarre; et ne parloient plus leurs lettres que de ceulx qui mouroient de fain chascun jour et que plus ne pouvoient tenir que dix jours, et puys huyt, et telle heure les veïs à trois...
          [Commynes, *Mém.*, VIII, 16 (III, p. 222)]

The phrase *de jour en jour*, frequently met with in Middle French,[89] expressed the idea of an indefinite number of stages in the course of events. Time appeared as a road with countless milestones which, like telegraph poles seen from a railway carriage, slipped monotonously by at unvarying intervals. Change was perceived as a succession of points. The typical medieval view of time, which was familiar only with sharply defined divisions and sudden changes, gave way to a new conception which everywhere saw transition and slow growth. Things were seen to grow and mature like plants.

Chastellain:
Avecques les jours leur croissoient plus et plus leurs misères.
          [II, p. 91]

Plus est venu le temps avant d'année en année, plus est multipliée en vous ceste erreur.   [V, p. 171]

Roi René:
De entendre a soy faire vouloir si que leur los et fasne ['fame']
puisse estre en croissant tousiours de bien en mieux.   [I, p. 56]

It is noteworthy that we now see the word 'grow' employed for the first time to denote gradual change, a usage still current today.

Since everything changed with time, and since the effects of an action and the light in which it appeared might vary according to the time at which it was done, it was a matter of importance for the would-be successful person to choose the right moment in which to act. The one who was alive to changing reality knew when conditions were favourable and sought out opportunities. It was a part of the art of living to be able to recognize the appropriate moment and turn it to account. This turn of thought was alien to the specifically medieval mentality. Neither the nobility nor the Church was familiar with the idea of the opportune moment. They desired no *tact des circonstances* and no departure from the norm in selecting the moment in which to act. However, a regard for the opportune moment that was more realistic and more in keeping with practical exigencies is discernible even in the most ancient French proverbs, which counselled a sober and solid wisdom.

Fol se targe et le terme approche.[90]
Fous vount à vespres e sages à matines.[91]
L'en deyt batre le fer tant que soit chaud.[92]
Qu mount plus tost q'il ne deyt chet plus tost q'il ne devereyt.[93]
Tutes hures ne sont meures.[94]

They spoke of the choice of the right moment without using the word *temps* itself. This was a way of thinking peculiar to Old French, popularly current in addition, and which disregarded the time factor even in those cases in which it would appear to us today as a matter of self-evident association. It was not until the later Middle Ages that the recognition and utilization of the appropriate moment became a conscious mental attitude. In that age, mental circumspection in the matter of the moment for taking action prevailed. Appropriate action became identified with the choice of a suitable time. Writers—historians in particular—were fond of describing this pondering over the proper time. After the murder of John the Fearless in 1419, Jean de

Thoisy and Athis de Brimeu were preoccupied with the question of the most opportune time to convey the news of the murder to Philip, who was then in Ghent.

> Sagement toutes-voies avisèrent de l'heure et de la manière plus expédient; et espièrent le temps que plus seroit en sa solitude avec ses prochains seulement, desquels ils désiroient plus l'assistence.[95]

It might be argued that this concern for the opportune moment is something that is met with at all times and was not a characteristic peculiar to the later Middle Ages. Here we observe not only the existence of a particular way of thinking, but also a novel evaluation of it. Only now, in the fourteenth and fifteenth centuries, did the choice of a time of action become a matter of such acute concern that it was considered necessary to mention and describe this concern. If, in the early Middle Ages, a moment or a day was selected with particular care, this was a consequence of a symbolic and formalistic mode of thought. Now it was the specific situation, the development of circumstances up to a certain degree of maturity, and external factors in general, that were the decisive factors in the choice of a time for action. This time was designated by a number of adjectives, which in Old French could not be used in this way.

Chastellain:

L'heure lui sembloit propre.	[I, p. 297]
L'heure y estoit convenable.	[I, p. 330]
Hors de heure.	[II, p. 252]
Selon la convenableté du temps.	[IV, p. 466]
En temps convenable.	[V, p. 142]
Heure toute apte et ydoine pour déjeuner.	[V, p. 258]
En temps opportun.	[VII, p. 234]

Monstrelet:
A quelque heure qui nous plaira et semblera mieux expédient.
[I, p. 48]

The implementation of the opportune moment, as did the purposeful use of time in general, became a political weapon in the fifteenth century. The realization that bravery and numerical superiority had to be backed by cunning and calculation ('cauteleux temporisement'[96]) if success was to be achieved, led to a greater regard for temporal possibilities and advantages. In the quarrel between Louis XI and Charles the Bold each was anxious to forestall the other and keep him wondering, to wait for an opportunity and to defeat his adversary with the aid of time:

> estoit une peur de prévention et de sens, pour à autruy rompre son entreprendre.[97]

In this art of enlisting time in the service of his political purposes the king was more adept than others.

> Et compassa fort bien son temps et faisoit une merveilleuse diligence; et avez bien entendu comme il dissimula à ce Symon de Quingy bien l'espace de huict jours, et que cependant advint ceste mort. Or sçavoit-il bien le dit duc desiroit tant la possession de ces deux villes qu'il ne l'oseroit courrousser et qu'il luy feroit couler doulcement quinze ou vingt jours, comme il fit, et que cependant il verroit ce qu'il y feroit.[98]

Commynes also now used the word *temporiser* to convey the anticipation of the favourable moment, and the deceptive delaying tactics of the opponent:

> Quantes sortes de gens luy en devindrent ennemys et se declairèrent, qui le jour de devant temporisoyent avecques luy et faignoyent amys!   [Mém., V, 1 (II, p. 105)]

> Sembloit que elle voulsist temporiser et commencer à reprendre quelque chose avec le roy.   [*Ibid.*, V, 2, (II, p. 114)]

The vision of the future

> Porveance est uns presens sens qui enquiert la venue des futures choses; ce est à dire que porveance est en II manieres et qu'ele a II offices: l'une est que ele pense, et remire les choses qui sont presentes; et par icele considere et voit devant toz ce qu'il en puet avenir et quelle puet estre la fin dou bien et du mal; et puis que ele a ce fait, si se conseille et se garnist par son savoir contre la mescheance qui avient. Por ce doit on devant veoir le mal qui avenir puet.   [*Trésor*, II, II, 54]

These words of Brunetto Latini indicate the direction to be taken by the relationship of humanity to the future as it followed his precepts. The anticipation of things to come was, like the careful weighing-up of the right moment for action, part of the adroit managing of one's life. This bourgeois and unheroic tendency to eliminate risk and hazard by dint of foresight was, it is true, here championed by an Italian who quoted Seneca, Gregory, Boethius, Tullius and Solomon in reinforcement of his arguments, and yet it seems to us, in the context of the French mentality of the later Middle Ages and of the consciousness of the Frenchman of the fourteenth and fifteenth centuries, as an attitude towards time that was peculiar to this age. The ideal to be striven for in the conduct of life was *prudentia*, an ideal which aimed at reducing the element of chance in action to a minimum and at bringing about a state of unremitting preparedness with regard to the future. To this end, all future possibilities had to be recognized and particular attention paid to those likely to materialize. Action that was subject to such extreme circumspection became ponderous and involved. Strategy took the place of adventure. The most remarkable literary manifestation of this new attitude towards the future was Chaucer's story of Melibeus,[99] with its image of the restrained and conscientious mode of life. The prudent, calculating spirit of the age found its most characteristic embodiment in the figure of Louis XI. Just as the uncertainty of the young person with regard to time is removed by the assurance with which the mature man deploys his time, so it was with the transition from the age of the Crusades to the circumspect mode of action of the later Middle Ages. Only now did one come to intimate terms with time: it appeared as a familiar medium; one was acquainted with temporal relationships; one could rely on time and wait, because one had already experimented with it. The future was synthesized from elements of the past. In the eyes of the men of that age, it lost its ethos and its qualities of mystery and unpredictability. It was seen as already coming into existence, developing out of the present; one was thus personally involved in this development. Everything was determined in advance, with some degree of verisimilitude.

Everywhere in literature, from the thirteenth century onwards, we find evidence of a vision of the future which regarded it as a determinable quantity. Quietly to bide one's time until the appropriate, predicted moment arrived, was one of the most noticeable attitudes of this age. The anticipation of things to come, even of those that were very remote, was often represented by Commynes as a personal idiosyncrasy and attitude of 'wise' contemporaries; this provided him with a new facet of humanity with which to work. The wide-ranging vision of the *personnages sages* often failed, in the turbulence of political affairs, to bear fruit:

> Vous voyez selon mon propoz tous ces seigneurs icy bien empeschéz; et avoient de tous costéz tant de saiges gens et qui veoient de si loing que leur vie n'estoit point suffisante à veoir la moytié des choses qu'ilz preveoient. [*Mém.*, III, 8 (I, p. 225)]

> Par especial à leur parlement... se trouvèrent plusieurs sages personnages et veoyent loing... [VI, 1 (II, p. 245)]

> Tant de personnaiges saiges, à qui la deffencé dudict royaulme touchoit, tant aliéz et soubstenuz, et qui voïent venir ce fetz sur eulx de si loing, qui jamais n'y sceürent pourvoir ne resister en nul lieu. [VII, 14 (III, p. 81)]

The thoughts of men increasingly roved beyond the bounds of the present; numerous mental associations were formed between what was now and what was to come; a present circumstance or action cast heavy shadows on the future. Everything that happened was thought of in terms of its probable consequences. Chastellain considered that the conduct of Louis XI during the early years of his reign would produce dire repercussions.[100]

The faculty now existed of seeing causes and effects, widely separated in time, as a whole. The ancient French spirit of impatience waned in the words and deeds of the men of this age, and faith in the future moderated the harshness of the present. Time would redress the balance and let justice be done. The historian knew that his contemporaries did not have the final say on historical circumstances and personalities.

> Le temps fera ses ennemis retourner à recongnoissance de leur évident tort.[101]
> Ung jour viendra qui payera pour tous.[102]

The qualities of 'patience' and 'espérance' thus assumed an important place in practical, action-orientated thought as in the moral and allegorical sphere of thought. The newly acquired capacity to see present events as the germ of future development was the cradle of historical thought. Today was causally bound up with tomorrow; the latter could be understood in the light of the former; the former foreshadowed the latter. The idea of succession acquired a new meaning: that of consequence. The succession of circumstances, formerly interpreted as a rhythm and experience, was now seen as a succession in a causal sense, as a chain in which each link was conceptually and factually connected with the other. The Middle French *s'ensuivir* had a shade of meaning different from that of *suivre*, which already existed. *S'ensuivir* was used chiefly to express the result and effect of a circumstance or state and was, when used with reference to the future, often tinged with the idea of that fateful course of events one was at pains to foresee.

> Monstrelet:
> Pour concorder iceulx princes et éviter le grant péril qui s'en povoit ensuivir. [I, p. 242]

> Pour éviter les inconvéniens qui se povoient ensuivir pour la cause de leur division. [I, p. 272]

The future is a true future only for so long as it is apprehended as something yet to come—that is to say, so long as it is expected, longed for or hoped for, desired or feared. A vision of the future free from all affective overtones is essentially identical with a sense of the present or the past. In Middle French that which was yet to come often lost its specifically future quality. The idea of 'hoping', which expressed a subjective interest in the future, was sometimes objectified, so that the word *espérer* was occasionally used in the sense of 'expect', 'anticipate':

> Comme contens ou descort fust ou esperast estre meu entre.[103]

> Monstrelet:
> Et de là fut ramené en ladicte ville de Mans, en son hostel, où il fut visité par notables médecins; néantmoins on y esperoit plus la mort que la vie. Mais par la grace de Dieu, il fut depuis en meilleur estat. [I, p. 8]

Interest in the future now appeared as a matter of contemplation. Formerly the language represented the future more particularly as something desired. Now the future was something on which one fixed one's eye. One perceived it clearly, or at least one wished to do so. Intention took the place of desires and demands in the minds of men. It would be misleading to conclude from what has been said that now only the cognitive faculties were turned towards the future and that desire played no part, as though one had no interest in life and the course of the world beyond a purely platonic one. No age was more concerned with the future, in every aspect of its mental and spiritual life, than this one. But, to us, its concern manifests itself primarily as a desire to know. Its will was not crippled, but sustained by a clear vision of the future. In contrast to what was said of the future in the *chanson de geste*, one now genuinely desired, in speaking of things to come, to adumbrate future events or individual conduct. There was a desire to indicate one's intentions for the future rather than one's present mental state. This more factual attitude was indicated by forms of expression such as the following.

Monstrelet:
J'ay délibéré de vous nommer et déclairer aucuns de voz ennemis et des miens. [II, p. 118]

Maistre Jehan Petit, docteur en théologie, lequel le duc d'Orléans avoit en propos de poursuyr et faire accuser de hérésie. [II, p. 123]

Henry de Lancastre... descendit au port de Touque en Normandie, en entencion de conquerre et mectre en son obéissance toute la duchie. [II, p. 188]

Les Anglois, véant que point ne seroient combatus, prinrent conseil et conclurent l'un avec l'autre de passer la rivière d'Oise se ilz povoient. [VI, p. 13]

This insight into the course of time brought with it the ability to wait. There was no necessity to make haste. Preparations, precautions and discussions assumed an important place in thought and life. Contemporary historians often attached especial significance to actions which had been the object of 'ripe deliberation' (*meure deliberation*)[104] or of 'lengthy preparation'.[105] The tendency to impulsiveness and impatience gave way to a new ethos, that of mature reflection. The elaborate preparations which preceded the events described by Monstrelet and Chastellain invested them with a weighty and enduring momentousness.

In the *Chanson de Roland* the future was felt as the goal and result of a will, and mainly of a human will. It was not thought

B

of as an objective process and necessity independent of the will. An objectivization of the future was implicit in the conception which saw one's future acts, not as the result of one's mental attitude and personal desires, but as the necessary and predictable consequence of a cause. In his poem *La Plainte des Pauvres Paysans de France* Monstrelet put the following prophecy, addressed to the king, on the lips of the peasantry:

Hélas! Très noble roy de France
Le pays de vostre obeissance
Espargnez le, pour Dieu mercy.
Des laboureurs ayez souvenance.
Tout avons pris en pacience
Et le prenons jusques icy.
Mais tenez vous assour, que si
Vous n'y mettez aucun remède,
Que vous n'aurez chasteau, ne ville;
Que tous seront mys a exille:
Dont jà sommes plus de cent mille,
Qui tous voulons tourner la bride.
Et vous lairrons tous esgaré;
Et pourrez cheoir en tel trespas
Qu'il vous fauldra cryer hélas!   [VI, p. 180]

Here the future—and indeed the future which was to be brought about by human agency—was seen as an external necessity. One saw what one proposed to do, not as something to be performed by oneself, but as an event which would take place, like a natural phenomenon, according to fixed laws. It was seen not as something willed but as an historical necessity. Here it was not the peasant who shook a warning finger (he was merely a tool) but the impending effects of an inevitable process.

A characteristic of the new spirit was a sense of the expediency of one's acts. These appeared, and were appraised, as stages on the path to the attainment of an end. It was now intention which set the seal on a deed or activity. Personal action thereby lost much of its intrinsic value as the expression of temperament or conviction, and as a thing which might be considered as good or bad in itself. One's gaze was now directed towards the future, and the attainment of a goal. There now existed a clearer idea than formerly of the course of events, which were connected with one another in as much as they tended towards an objective and served a purpose. The idea of the goal to be reached assumed a place of noteworthy importance in the language of the Middle French period. Everything appeared to hinge on this idea. The saying *Au commencement pense à la fin* originated in the thirteenth century. The true import of *la fin* scarcely manifested itself in the age of Old French. At that time the end was not a thing that was striven for with such fixity of purpose as now. It appeared less as the final term of a series, as the fulfilment of effort, than as an event in itself, variously apprehended according to the situation existing at the time. Now, however, the thought of the goal, of the end that was desired and striven for, to the attainment of which all else was subordinated, manifested itself in language in the form of new phrases.

The narratives of the fifteenth century contained many such expressions of striving, purposefulness and mindfulness of the desired goal. We quote a few examples from the *Cent Nouvelles Nouvelles*:

venir à chef; venir à bout; mener à fin; estre à point   [73]

venir aux fins et intencions où il entendoit   [87]

à fin conduire   [10]

Le bon clerc emprint sur lui de la trèsbien conduire et à sa seure fin terminer   [13]

foison d'aultres raisons... tendans à fin de l'oster de son propos [13]

après la trèsdesirée conclusion de sa haulte entreprinse   [3]

The novel use of the word *la fin* in *afin de* and *afin que* reflected this new spirit of purposefulness. The element of intention bound up with the end to be attained assumed a place so prominent that the end lost its temporal significance. In Remy Belleau's comedy *La Reconnue* the neighbour said to the lawyer's wife, the latter having informed her of a decision taken by her husband, 'J'en crains une fin',[106] meaning 'I fear a ruse, a hidden intention'. The desired end, which lay in the future, became, in its verbal meaning, completely indistinguishable from present intention.

It will be apparent from what has been said of this concern for the future that men now took their destiny into their own hands with more determination than in the golden age of chivalry and monasticism. The realization that the future might be influenced and shaped in the present became a powerful force. Whereas the person imbued with the stern spirit of the Middle Ages associated the words 'future' and 'end' with death, the end of the world and the Beyond, the thought of what was to come now took on a more worldly character. The watchword was now to bestir oneself in the world, to plan, to provide, to take precautions and patiently to wait. The enhanced value attached to the life here below engendered a greater degree of concern for what happened in the world and for how one fared oneself. Although the world was regarded as a vale of tears even in the later Middle Ages, one nevertheless clung to it and desired to retain a permanent place in it. To the men of this age, it was prosperity[107] and long life that appeared as the most desirable ends; their concern for the future was motivated by these desiderata. The endeavour to eliminate the unforeseen and remove the element of chance from life was actuated by this ideal. In the struggle against hostile forces of all kinds—sickness, death, poverty, misfortune, defeat in battle, and the like—one strove to acquire a feeling of security and preparedness by incorporating the future into present thought.

The mental attitude of the person who intended to face the future rationally and systematically gained slowly and steadily in clarity and breadth in proportion as scientific thought and the recognition of natural laws advanced as the age of the Renaissance was approached. One way in which this desire to guard against future contingencies manifested itself was an enhanced awareness of the tasks to be fulfilled by medicine. It was the business of medical science to foresee as far as possible the course of a disease in its individual stages ('la parfaite connoissance et gradation des temps de la maladie'[108]); only scientific knowledge could make this possible. The art of healing must be released from the blind alley of conjecture and guesswork by 'methodical' and 'rational' procedures, as Ambroise Paré pointed out, quoting from the writings of a Paris surgeon, Thierry de Héry.[108] The greatest feasible number of *Zeitgestalten*—to use a term of modern psychology—must be created, so that the present might be linked with the future. Although the subjective pleasure in ephemera characteristic of the Renaissance was soon to wane, the matter-of-fact taking into account of future possibilities went on steadily gaining ground, and continued to do so up to the present day.

The value of time

The preoccupation with the future which we have noted may also be interpreted as an expression and consequence of the new importance acquired by time. As it was now felt as a positive force, and evaluated as such, greater attention was now paid to that part of time which was still to be lived through, employed, and regulated—in short, that part over which one had the greatest degree of control: the future. The past was now outside such control, and was thus worthless. It did, however, have its indirect usefulness in the shaping of life, in that the experience it provided was of service in determining the most profitable use of the time still at one's disposal.

The value of time now manifested itself in every aspect of life.[109] It was highly prized, not for its own sake, and not as a form of inner riches or inalienable property, but as a consequence of interested involvement in the course of the world and of the realization of the importance of the proper use of time. The 'medieval tendency to over-estimate the value of a thing or circumstance in itself', of which Huizinga speaks,[110] did not manifest itself in the view of time prevailing in the fourteenth and fifteenth centuries. Time was valued only in so far as it was used. There was no thought of prizing it for the sake of its unique qualities. It derived its value from its relationship to the other 'commodities' of life. Its value depending on worldly goods and happiness, it was no imponderable. The idea of time was now closely associated with such desiderata as advantage, usefulness, pleasure and power. The thought of saving, wasting and having time exercised the popular consciousness to a great extent. In 1464 Louis XI introduced a regular postal service with relay stations four miles apart.[111] The same century saw the invention of the printing press and the discovery of America, innovations towards whose realization the thought of saving time had also contributed. The good king René called it 'le temps desiré de tous'.[112]

The mentality from which this desire to gain time proceeded was far removed from the sense of values which impelled Seneca[113] to advocate strict economy of time. For him, time was the most precious of all possessions, one which must on no account be dissipated, and that in a much stricter sense than was understood by the medieval mind. Seneca saw time as a noble vessel of which only the most valuable contents were worthy. The idea of gaining and wasting time entertained by the Frenchman of the fourteenth and fifteenth centuries was bound up with pleasurable and useful activity, an attitude which Seneca despised. Time was a consciously used instrument in the daily struggle of life. Louis XI knew that time was power, and accordingly he set up an information service which afforded him the advantage of keeping abreast of his rivals and adversaries in the matter of time. Time could no longer be neglected; such neglect had to be paid for. Not only were things seen to clash with one another in space, but unfortunate and unprofitable combinations and coincidences were observed to occur in time also.

Time was only one of the many sources of power which were now turned to account. Thus a new European spirit came into being, which strove to exploit all means to the acquisition of power and to bring about the levelling of values. This spirit subsequently led to the coining of the maxim 'Time is money'; those who credited it believed they were exalting time, whereas in fact they degraded it. According to certain parochial reports originating at various times and places in the later Middle Ages, a new practice, typical of the changed attitude to time, became current: the hands of public clocks were moved in order to outwit the enemy. Even if the historical authenticity of such reports is open to doubt, their documentary value in revealing the mentality of the age still stands. Such devices had meaning only in a world which depended on the clock to a great extent, and in a society which was largely regulated by the pealing of bells, and which therefore could be injured, influenced or deceived by interference with them. For the person who thus tampered with official time, however, such practices were in fact a means to an end valued more highly than time itself. Time was no longer unchangeable and inviolate; it no longer belonged to God, but to man. The medieval respect for the temporal dimension as an autonomous domain with its own laws was in decline. Time was now ruthlessly manhandled. The knight of the palmy days of the Middle Ages declined to avail himself of all the advantages at his disposal in combat with his adversary. Now all drops were collected, the cloudy and evil-smelling along with the pure and clear; they needed only to swell and accelerate the river of personal interest. Things and even persons were divested of their intrinsic value.

In the language of the fifteenth century new turns of phrase appeared which threw a characteristic light on the current manner of assessing and handling individuals. One spoke of 'winning', 'exploiting', 'attracting', 'making use of' and even of 'buying' a person.[114] To one who was concerned with the exploitation of all possible resources, others appeared as mere tools to be used in order to gain an end. This contempt for the intrinsic worth of men and things went hand in hand with a neglect of time as an experience. Just as one could 'gain' a person, so one could also gain time. There was no way of expressing this idea in Old French. In *La Vie de Guillaume* of Chrestien de Troyes we find this example of a desire to gain time: when the unknown queen, wooed by the aged Gleolai, asked him that he forgo his connubial rights for one year, she did so in the hope that time would enable her to escape from her difficult situation. The characters created by the courtly poet thus behaved as people who knew the value of time, but several centuries were still to pass before the appreciation of its value became a definite and universal concept. The postponement requested by Gratiana was an isolated instance of its kind in the twelfth century. The idea of gaining time had become widespread amongst the historians of the fifteenth century.

Monstrelet:
Mais ce fut une décepcion que ledit prestre faisoit afin de les delaier et atarger de paroles tandis que les Anglois s'assembleroient pour les venir combatre.   [I, p. 92]

Chastellain:
Il sembloit à ces seigneurs dauphinois... qu'il seroit bon,... que ils besongnassent et exploitassent temps à bon profit.   [I, p. 228]

Et par ainsi vouloit laisser couler leur première aigreur, pensant par intervalle de temps après pouvoir vaincre.   [III, p. 458]

Il y falloit bien entendre et tourner temps pour tout y mettre en
point.   [V, p. 361]

Commynes:
Il le faisoit seullement pour leur donner occasion de parler ensemble
et de gaigner temps, car ilz avoient de coustume, et ont encores,
d'aller tout le peuple ensemble au palais de l'evesque...
                                                    [II, 3 (I, p. 112)]

Et encores quelque congnoissance qu'ilz eussent que le roy, nostre
maistre, le feïst pour gaigner temps et faire son faict en ceste
guerre, qu'il avoit commencée, si le dissimuloyent-ilz pour les
grans prouffitz qu'ils en avoyent.   [VI, 1 (II, p. 242)]

Ainsi, sur ces dissimulations, ung moys ou deux de terme gaigné en
allant et en venant, est rompue à son ennemy une saison de luy
mal faire.   [VI, 1 (II, p. 242)]

Examples like these make it evident what was achieved by
gaining time. A clear idea of the value of time is also apparent.
To the active and opportunist spirit, time was a thing to be
reckoned with, a material to be expended to the best advantage.
It was imperative to avoid loss of time. The individual was
thought of as the master of time; he deployed it in his utilitarian
wisdom. To waste time was a blunder which had to be paid for by
defeat in the battle of life. Jean de Meung, with his practical
and earthy view of life, likewise warned of the dangers of
wasting time:

Roman de la Rose:
Car tant pert de son tens, la lasse!
Cum sens joir d'amors en passe.   [v. 14426]

En la fin encor le sauras
Quant ton tens perdu i auras
Et dégastée ta jonesce
En ceste dolente léesce.   [v. 5338ff.]

The phrase *perdre son tenz*, however, is to be found as early
as the twelfth century.[115]

Cligès:
Car se mil anz avoie a vivre
Et chascun jor doblast mes sans,
Si perdroie je tot mon tans.   [v. 2738]

It did not have the sense, familiar to us, of 'wasting time'. It will
be remembered that, in Old French, the concepts of 'time' and

'life' were more closely linked than in later ages. The author of
the *Chanson de Roland* often used *tenz* in the sense of 'life'. This
meaning is apparent in *perdre son tenz*, for in using it one thought
of that part of one's lifetime still at one's disposal. To waste one's
time was the same as committing one's life to something worthless and useless. This meant more than wasting time in the sense
of missing an opportunity. It meant sacrificing part of one's soul,
discarding part of one's being. The ancient phrase lived on into
Middle French, and was complemented by another, *perdre temps*.

Deschamps:
Perdre temps et avoir. [I, p. 110]

Du temps perdu a no destruction. [II, p. 17]

François perdent leur temps a conseillier. [II, p. 91]

Je plain et plour le temps que j'ay perdu. [II, p. 93]

Christine de Pisan:
Par ce que plus ne s'en tendroient aux pertes de temps que faire
souloient, ains chacun a son droit mestier.

[*Livre de la Paix*, 3, 15]

Monstrelet:
Ceulx de la Bastille voians qu'ilz perdoient leur temps de la tenir.

[III, p. 266]

Cent Nouvelles Nouvelles:
Vous perdez temps. [88]

Ledit maistre d'ostel perdoit son temps, car, quelque chose qu'il
sceut remonstrer, si ne la voulut il croire. [99]

Commynes:
...de ce que les Bourguygnons s'estoient mys à pied, et puis
remontéz à cheval, leur porta grant perte de temps et de dommaige.

[I, 3 (I, p. 24)]

La moytié du temps se pert avant qu'il y ait rien conclud ne accordé.

[I, 16 (I, p. 91)]

Le roy, qui estoit plus saige a conduyre telz traictéz que nul autre
prince qui ait esté de son temps, veoit qu'il perdoit temps s'il
ne gaignoit ceulx qui avoient le credit avecques son frère.

[II, 15 (I, p. 170)]

Fault parler du duc d'Orleans, qui, quant il eut pris le chasteau de
Novarre, y perdit temps aulcuns jours, et puis tira vers Vigesve.

[VIII, 6 (III, p. 157)]

Recueil de Soties:

Vous fault il qu'on temps perde?   [II, p. 48]

Or temps perdons.   [II, p. 55]

On perdroit temps.   [II, p. 66]

The sense of *perdre son temps* is not always clearly distinguishable from that of *perdre temps*, but the latter phrase indicated an aspect different from the twelfth-century conception of time. 'To lose time' now had the literal meaning of wasting a certain amount of time one might use to better advantage. The alternatives implicit in the Old French form were 'to devote, or not devote, one's time to something'. The thought underlying the Middle French phrase was one of a smaller or greater degree of loss of time. It was all a matter of quantity: time was a stuff which had to be used to the best advantage. Chastellain said praisingly of Charles the Bold that he 'perdoit peu d'heures'.[116] The less time was valued as an experience (its inherent value) the more one was convinced of the worth of time as such.

The belief in the possibility of recovering it provides further evidence of its being treated as a thing. Guillaume de Machaut narrated the story of a love between a poet of sixty and a girl of eighteen. The following sentence occurred in a letter sent to the girl by the old man:

> Menons si bonne vie que nous porrons, en lieu et en temps, que nous recompensons le temps que nous avons perdu.[117]

One believed one was respecting time, whereas in reality it was degraded. It had altogether become the object of the human desire to live and to gain power. It was no longer within man, nor over him, but before him—at his disposal to use as he thought fit. He was master of it. He knew the importance of saving time:

> N'y espargnant le temps pour le travail par moy faict nuit et jour.
> [Paré, I, 2, 3]

The task imposed on man by time did not consist in apprehending duration as something unique, and in consciously living through it, but in the rational deployment of the quota of time allotted to him. The extravagant and effervescent vitality of the later Gothic era was curbed by the new way of thought, which reckoned in terms of time.

This awareness of the value of time also found its expression in the manner in which one distinguished between the essential and the incidental in conversation and story-telling; the incidental was omitted or cut short. One was disinclined to waste one's own time or to make unnecessary demands on that of the listener. In the fifteenth century such phrases as 'in short', 'in a few words', 'I pass over that', 'to cut a long story short', 'we cannot go into details', and the like, became frequent. They were evidence, less of the impatience of an ebullient temperament, than of the capacity of an enlightened understanding to make the best use of the time at one's disposal for telling a story, or conducting a conversation or business, by concentrating on essentials and keeping in mind the necessity for speed.

Fabliau de Sire Hain et Dame:
Que vous feroie plus lonc conte?   [Montaiglon, I, p. 102]

Charles d'Orléans:
Pour plus abbreviacion.[118]

Monstrelet:
Veuillez abréger le temps de en mander vostre plaisir.   [I, p. 45]

Chastellain:
J'en diffère le réciter pour cause de brièveté   [I, p. 111]

Cent Nouvelles Nouvelles:
Je passe en bref.   [1]

Pour trousser le compte court.   [1]

Pour abregier.   [3]

En brefs mots.   [5]

Pluseurs aultres mistères trop longs à racompter.   [2]

Sans faire long procès.   [22]

Que pour abreger je trespasse.   [26]

Si m'en passe pour cause de brefté.   [60]

Pathelin:
Bref je suis gros de ceste piece.   [v. 220]

A brief vous dire.   [v. 794]

Commynes:
D'autres chefz y avoit-il, que je ne nommeray pas pour ceste heure
pour briefveté.   [I, 2 (I, p. 12)]

Et, pour le vous faire court, il sejourna aucuns jours en la cité.
[II, 4 (I, p. 117)]

Pour abrevier ce propos.   [VII, 2 (III, p. 16)]

Toutesfoiz, pour abreger, le saul-conduyt fut accordé et envoyé.
[VIII, 16 (III, p. 231)]

Des Périers wrote a short story caricaturing this tendency, in which a monk replied monosyllabically to each question put to him, in order to avoid wasting time in speaking while at table:

'Quel habit portez-vous?' — 'Froc.' — 'Combien estes-vous de moynes?' — 'Trop.' — 'Quel pain mangez-vous?' — 'Bis.' — 'Quel vin bevez-vous?' — 'Gris.' — 'Quelle chair mangez-vous?' — 'Beuf.' — 'Combien avez-vous de novices?' — 'Neuf.' — 'Que vous semble de ce vin?' — 'Bon.' — 'Vous n'en bevez pas de tel?' — 'Non.' — 'Et que mangez-vous les vendredy?' — 'Œufs.' — 'Combien en avez-vous chascun?' — 'Deux.'   [II, p. 207 (Nov. 58)]

Here language was treated as a quantity, exactly as was time. Words were considered as units of duration. Thus language became not only a means of expressing the new ideal of economy of time, but also a means of giving effect to it in life. Language, and the hearing or reading of it, was a process and therefore occupied time. Speech, like external time, was a measurable entity. Economy of words meant economy of time.

The desire to gain time had its much more serious side, manifesting itself in the awareness of the brevity of human life and the wish to defer the coming of death, the door to eternity. A mental attitude unknown to the chivalric and medieval spirit was born, an attitude which clutched apprehensively to the earthly life that fled away so swiftly.

Le temps de l'espace de vie, dont il faut à nous tous rendre compte, lequel se passe courant incessamment comme eaue de riviere sans se arrester et va sans revenir,   [IV, p. 2]

lamented the good king René of Anjou. The active person must sedulously avoid wasting time. One's lifetime was a gift of God, the use of which would have to be accounted for. In the *Mortification de la Passion Vaine* of King René, Fear of God said to the Soul:

Pense à ton fait sans plus perdre de temps.   [IV, p. 9]

It was to be hoped that Christ would prevent the appearance of the devil as one lay dying:

> Pécheur, tu es à moy;
> Te souviengne de ton desroy
> Et du temps que tu as perdu,
> Car tu es à fin venu.[119]

Everyone must needs die, but no one knew the hour of his death. This question haunted the medieval mind. We must all die, 'Mais quant? Là gist le point.' If fools knew this, René observed, they would divide their lives into two halves, attending to mundane matters in the one and serving God in the other.[120] But this mathematical way was impossible. One did not know how much time one still had left. And the fool did not know how to make use of his time. He suffered the same fate as L'Abuzé en Court in the allegory of that name. A young man arrived at court and was joined by Le Temps, 'a fair youth of strange appearance'. He now had time in his service, but in his life at court he forgot to make good use of it, failing to keep pace with it as it fled away, and after twenty years he became painfully aware of the fact that he had lost it. He had kept company with the allegorical personality, Abuz, instead of remaining faithful to time. 'How did I come to lose it?' he exclaimed when Abuz told him the truth, which he was reluctant to accept. After reflecting a little, however, he realized that he had been cheated of time. The fair words and promises spoken to him at court had made him waste it.

> J'ay perdu temps et richesse,
> Toute joye et esbatement,
> Force, beaulte, sens et jeunesse,
> Pour croire trop legierement.[121]

The fool would gladly recover lost time, but that he cannot do, for another has seized on it:

> Si le temps pers, un autre le remeuvre;
> Si tu le quiers, un aultre l'a trouvay;
> Quant folioys, un autre en science euvre,
> Quant plus ne l'as, aultruy l'a recouvré;
> Qui a le temps doibt estre bien gardé,
> Et qui ne l'a, à l'avoir painne mecte.[121]

This was the high-water mark of the wisdom of the century in the matter of time. The urgency and didactic breadth with which the aged René treated the theme of lost time is evidence of the importance that was attached to this question. To have time at one's disposal was a requisite of practical competence in life as well as an ideal of religious understanding, and the one who wasted his time injured both his mundane interests and his chances of happiness in the life to come in equal measure. Thus the idea of wasting time lost its objective quality and its element of moral and religious indifference.

Notes

1 Christine de Pisan (ed. Roy, Paris, 1886), *Le Dit de la Rose*, v. 28; *Le Livre du Dit de Poissy*, vv. 34f. Eustache Deschamps (ed. Raynaud, Paris, 1878–1903), *Miroir de Mariage*, vv. 11482f., 11510f., 11785f., 11893f.; *Sur la Naissance de Louis de France* (II, 48).

2 Nov. 89 (ed. Wright, Paris, 1858).

3 J. M. Buffin, *Remarques sur les Moyens d'expression de la durée et du temps en français* (Paris, 1925), p. 99.

4 Cf. Uwe Ruberg, *Raum und Zeit im Prosa-Lancelot* (Munich, 1965), pp. 105–64.

5 Enguerrand de Monstrelet, *Chroniques* (ed. Douet d'Arq, Paris, 1857–62).

6 In this and all subsequent quotations from Commynes the first figures indicate the book and chapter of the *Mémoires*, the second the volume and page of Calmette's edition.

7 In this matter cf. Lucien Febvre, *Le Problème de l'incroyance au XVIe siècle* (Paris, 1962). The chapter entitled 'Temps flottant, temps dormant' (pp. 426–34) shows how far temporal ordering in everyday reality lagged behind the feeling for time as manifested in literature.

8 *L'Hystoire et playsante Cronique du Petit Jehan de Saintré* (ed. Guichard, Paris, 1843), ch. 26.

9 *Ibid.*, ch. 46.

10 *Ibid.*, ch. 73.

11 *Recueil de Farces, Soties et Moralités* (ed. Jacob, Paris, 1876), p. 182.

12 *Œuvres du Roi René* (ed. Quatrebarbes, Angers, 1845).

13 The attribution of dates is based essentially on Gamillscheg, *Etymol. Wörterbuch der franz. Sprache* (second edition, Heidelberg, 1966) onwards.

14 Most of these acquisitions of the twelfth century occur in Philip's *Cumpoz.* See p. 57.
15 Deschamps, VII, p. 274.
16 *Ibid.*, IX, p. 266.
17 *Ibid.*, IX, p. 260.
18 Villon, *Gr. T.*, v. 795.
19 *Condamnacion de Bancquet*, Jacob, p. 402.
20 Monstrelet, I, p. 6.
21 Paré, II, p. 527.
22 Ronsard, I, p. 70.
23 Brunetto Latini, *Li Livres dou Trésor* (ed. Chabaille, Paris, 1863), II, ɪɪ, 14.
24 Chastellain, V, p. 250.
25 *Ibid.*, I, p. 234.
26 *Ibid.*, V, p. 101.
27 *Ibid.*, III, p. 362.
28 *Cent Nouvelles Nouvelles*, 15.
29 Chastellain, III, p. 298.
30 *Ibid.*, IV, p. 215.
31 *Ibid.*, VII, p. 107.
32 *Cent Nouvelles Nouvelles*, 100.
33 Chastellain, V, p. 459.
34 Paré, II, p. 630.
35 Rabelais, IV, 44.
36 Montaigne, *Essais*, 'De la douleur'.
37 Ronsard, VII, p. 84.
38 *Ibid.*, Ode 22.
39 Marot, III, p. 145.
40 Du Bellay, I, p. 246.
41 Ronsard, Ode 19.
42 Montaigne, *Essais*, II, 29.
43 French examples are to be found in Antonín Šesták, *Pojem času v jazyce francouzském. La notion du temps en français [Sur les traces de la notion 'espace–temps' dans les langues romanes]* (Brno, 1936).
44 P. Imbs, *Les Propositions temporelles*.
45 Kurt Pfister, *Die mittelalterliche Buchmalerei des Abendlandes*, (Munich, 1922), p. 31.
46 Brunot, *op. cit.*, I, pp. 466f.
47 M. Roy, I, p. 119.
48 *Œuvres*, I, p. 81.
49 Montaiglon–Raynaud, *op. cit.*, I, p. 80.
50 Jacob, *Farces*, p. 425.
51 *Les Quatre Ages de l'homme* (ed. Fréville, Paris, 1888), p. 1.
52 Bonaventure des Périers, *Œuvres françaises* (ed. Lacour, Paris, 1856).

53 Halle, 1914; reprinted Tübingen, 1966.

54 Tobler–Lommatzsch's *Altfranzösisches Wörterbuch*, under the head-word *entre*, cites only one example relating to the concept of 'between two points in time': 'Entre deus samedis avient moult de mervoilles'—*Prov. frç.*, M. 691.

55 Monstrelet, V, p. 158: 'sans intercupacion'.

56 Roi René, IV, p. 75.

57 Montaiglon–Raynaud, I, p. 172.

58 Another word of originally spatial meaning is *environ*, which occurs in temporal use in the fourteenth century. See K. Ringenson, '*Il a dans les cinquante ans*', *Studier i Modern Språkvetenskap*, 14, p. 140 (Uppsala, 1940).

59 Paul Imbs, 'La Journée dans la Queste del Saint Graal et la Mort le Roi Artu,' *Melanges de philol. romane et de litt. médiévale offerts à E. Hoepffner* (Paris, 1949), pp. 279–93.

60 *Op. cit.*, p. 290.

61 *Op. cit.*, p. 291.

62 *Libri Psalmorum versio antiqua Gallica* (ed. Michel, Oxford, 1860).

63 Tailliar, *Recueil d'Actes des XIIe et XIIIe siècles en langue wallonne* (Douai, 1849).

64 *Aufsätze zur rom. Synt. u. Stil.*, 'Noctem et diem', p. 278.

65 Bartsch, *Rom. u. Past.*, p. 44.

66 *Op. cit.*, p. 48.

67 *Op. cit.*, p. 51.

68 *Op. cit.*, p. 73.

69 J. Trénel, *L'Ancien Testament et la langue française du moyen-âge* (Paris, 1904), p. 231.

70 144.

71 187.

72 Spitzer, 'Aujourd'hui', *Ztschr. f. rom. Phil.*, 50, pp. 336f., considers *le jour d'hui* as an abbreviation of *le jor qui est hui*.

73 *Op. cit.*, p. 339.

74 Le Proux, *Chartes françaises du Vermandois de 1218 à 1250*, Bibl. de l'École des Chartes, XXXV.

75 Wilmotte, 'Études de Dialectologie wallonne', *Romania*, XVII.

76 Spitzer, *op. cit.*, p. 337.

77 Spitzer, *op. cit.*, thinks that *jour d'hui* is opposed to *jour d'hier* and *jour de demain*.

78 *Œuvres*, V, pp. 397f.

79 *Ibid.*, V, 402.

80 *Ibid.*, V, 492.

81 Anonymous French translation of Alain Chartier's *Dialogus Familiaris*, ed. G. Rosenthal, *Jahresbericht der Klosterschule Rossleben*, 1911–12, p. xviii.

82 Froissart, *Poésies* (ed. Scheler, Brussels, 1871), II, p. 375.
83 *Le Roman de Jehan de Paris* (ed. Mabille, Paris, 1855), p. 49.
84 Cf. S. Heinimann, *Das Abstraktum in der französischen Literatursprache des Mittelalters* (Berne, 1963), pp. 169f.
85 Chastellain, VII, p. 1.
86 *Ibid.*, VII, p. 220.
87 *Farce de Mestier et Marchandise* in M. E. Fournier, *Mystères, Moralités, Farces* (Paris, 1872), p. 47.
88 Rosenthal, *op. cit.*, pp. xx–xxi.
89 Cf. Deschamps, I, p. 119; II, pp. 68, 115, 126, 226; III, pp. 26, 109; etc.
90 Le Roux de Lincy, *Le Livre des Proverbes français* (second edition, Paris, 1859), II, p. 476.
91 *Ibid.*, II, p. 476.
92 *Ibid.*, II, p. 477.
93 *Ibid.*, II, p. 481.
94 *Ibid.*, II, p. 483.
95 Chastellain, I, p. 43.
96 *Ibid.*, V, p. 488.
97 *Ibid.*, V, p. 459.
98 Commynes, *Mém.*, II, 3 (I, p. 231).
99 *Canterbury Tales* (ed. Koch, Heidelberg, 1915), vv. 13871–14671.
100 Chastellain, V, p. 13.
101 *Op. cit.*, VII, p. 233.
102 *Saintré*, ch. 14.
103 1313, Godefroy, *Dictionnaire de l'ancienne langue française*, quoted from H. Hatzfeld, *Die Objektivierung subjektiver Begriffe im Mittelfranzösischen* (Munich, 1915), p. 84.
104 The expression occurs in Chastellain, III, p. 311; IV, p. 361; Monstrelet, I, p. 23; VI, p. 166; Roi René, I, p. 106.
105 Chastellain, V, p. 481. *L'Escoufle* (ed. Michelant–Meyer), V, p. 576, quoted by F. Brunot, *op. cit.*, I, p. 288.
106 R. Belleau, *Œuvres complètes* (ed. Gouverneur, Paris, 1867), III, p. 328 (Act IV, scene 3).
107 H. Hatzfeld, 'Der Geist der Spätgotik in mittelfranzösischen Literaturdenkmälern', *Idealist. Neuphilologie* (Heidelberg, 1922), p. 206.
108 Ambroise Paré, *Œuvres complètes* (ed. Malgaigne, Paris, 1840–1841), II, p. 545.
109 The first writer to draw attention to the value now attached to time was Alexander von Gleichen-Russwurm in *Die gotische Welt* (Stuttgart, 1919), p. 5.
110 J. Huizinga, *Herbst des Mittelalters*, German trans. T. Wolff-Mönckeberg, third edition, Munich, 1931, p. 347. (Published in

Great Britain by Penguin Books under the title *The Waning of the Middle Ages*.)

111 Funck–Brentano, *L'Histoire de France racontée à tous: Le Moyen-Age* (Paris, Hachette), p. 511.

112 *Œuvres*, IV, p. 99.

113 *De brevitate vitae*.

114 Cf. R. Glasser, 'Zur Objektivierung menschlicher Beziehungen im französ. Spätmittelalter', *Archiv für das Stud. der neueren Sprachen*, 173, pp. 202–8.

115 We do not number 'Maint gentil home perdi iluec son tens' (*Aliscans*, v. 216) amongst these examples, for it has not the same significance. It means 'lost his life thereby'.

116 *Œuvres*, VI, p. 229.

117 G. de Machaut, *Le Livre du Voir-Dit* (ed. Paris, 1875), p. 203.

118 *Poésies* (ed. Champion, Paris, 1923), p. 161.

119 Chastellain, VI, 61 (*Le Pas de la Mort*).

120 *Œuvres*, IV, p. 14.

121 *Ibid.*, IV, p. 139.

4

The concept of time in the Renaissance

The new subjectivism

If we consider the linguistic, literary and other cultural manifestations of the thirteenth, fourteenth and fifteenth centuries not only, as we have so far done, as the expression of the consciousness of time in the later Middle Ages, but also as a source of developments leading to present-day thought, we shall realize that this epoch made a greater contribution than any other towards forming a view of time which still has its adherents today. The spatial and external concept of time, which first found its decisive and conscious expression in Middle French, is of permanent value. The civilizing element inherent in the objective view of time remained in the language, and hence in mental concepts, and was therefore perpetuated, even as the popular civilization set up in the later Middle Ages was preserved, despite numerous changes and reactions. Although a strong emphasis on time considered as an experience is discernible during the Renaissance and the Romantic period, this did not drive out the sense of clarity and direction which had informed the popular consciousness, but actually added to the possibilities inherent in the human attitude towards time. The process of clarification of the temporal image of the world which we have followed up to this stage of our inquiry was continued during the sixteenth century and received a stimulus from the humanists' preoccupation with the temporally remote world of cultural antiquity, as also from the increasing vigour of scientific methods of thought.

The most sensitive and original minds, however, began to view the objective temporal order as a constraint and a confinement of the individual feeling for life. It is with their view of time that we must now concern ourselves. We shall endeavour to present it as a coherent and organic reorientation of the human attitude to time, despite the many differences in the outlook of its adherents and in their spheres of activity. The changed attitude of man towards time did not become apparent until the latter half of the sixteenth century, when it manifested itself in the work of the Pléiade and of Rabelais and Montaigne. For them, time had become an object of incessant inner preoccupation, with regard to which they desired to take up a definite standpoint and exercise a firm individual will.

Between them and the Middle Ages stood Clément Marot, who had nothing to say about time, and whose poetry showed no trace of those elements which we see as typifying the view of time current in the Renaissance. Marot had not that active attitude to time shared by his successors, who saw time as something extraneous, as a duty and responsibility. Marot was at one with time; it was within him, and hence he did not become aware of it and did not feel the need of coming to terms with it. His subjectivism was not willed; he did not define and propagate his art of living, and his impatience did not clash with the rigid framework of time.

The view of time of the poets properly of the Renaissance was a matter of desire and longing, more an ideal than its fulfilment. They were too much at odds with time and thus showed that the inner harmony of man with time was something to be striven for rather than something actually existing. For Rabelais it remained an ideal, and Montaigne experimented and juggled with time without attaining that state of rest which Wölfflin considered to be a characteristic of the spirit of the Italian Renaissance. 'In its perfect creations, we find no trace of constraint, restlessness or excitement.'[1] The French placed too much stress on intentions, slogans and programmes in their literary manifestations. Their relationship to time lacked that self-evident repose which informed the songs of Ariosto. These are the reservations with which, we believe, any account of the new attitude to time must be prefaced.

Time did not affect everyone in the same way. It turned a different face at different times and to different individuals. The experience of objective entities deprives them of their fixed quantitative value. The same time span which yesterday seemed to me long may seem short tomorrow. Time cannot be equated with the clock. For the impatient one an hour is an eternity, while for the happy man whole years pass rapidly. This truism was an old adage, which had been handed down to the Middle Ages from various sources. Its use can be traced via the Provençal poets and the Old French epics influenced by antiquity—'Plus d'un an a ore en un jor'[2]—to the Italians, who made of it a standing formula of impatient expectation.[3] The Renaissance poets now conveyed this experience with a new power of conviction. Indeed, Marot made witty use of the paradox, playfully disguised as the expression of the art of womanly exaggeration:

> Des que mamie est un jour sans me voir,
> Elle me dit, que j'en ay tardé quatre:
> Tardans deus jours, elle dit ne m'avoir
> Veu de quatorze, et n'en veult rien rabatre.[4]

But the later poets of the Renaissance shaped the theme more seriously and stripped it of its incidental quality, which had led to its use as a formula:

Rabelais:
> Dont nos espritz, taincts de merencolie,
> Par longue attente et vehement desir,
> Sont de leurs lieux esquelz souloient gesir,
> Tant deslochés et haultement raviz
> Que nous cuidons et si nous est advis
> Qu'heures sont jours, et jours plaines années.
> Et siecle entier ces neuf ou dix journees.[5]

Ronsard:
> Ces longues nuicts d'hyver, où la Lune ocieuse
> Tourne si lentement son char tout à l'entour,
> Où le coq si tardif nous annonce le jour,
> Où la nuict est annee à l'ame soucieuse.[6]

> Il me semble que la journée
> Dure plus longue qu'une année
> Quand par malheur je n'ay ce bien
> De voir la grand'beauté de celle,
> Qui tient mon coeur.[7]

Belleau:
> Toutes les heures me sont iours
> Si ie ne voy nostre voisine.

[*La Reconnue*, III, v. 327]

Des Périers:
Et luy dura la nuict plus de mil ans qu'il n'estoit desjà après nos
vengeances.   [II, p. 49 (Nov. 9)]

The language, in this century, was rich in resources with
which to negate objective duration, to represent it as inessential
and lacking in verisimilitude, and to convey only the impression
it made on the agitated soul. External time was nothing, experience was everything. These forms of expression did not indicate
an inability to conceive of, and live within, a set and rigid temporal
order, but they denoted a consciously mannered gesture of revolt
against time, which makes it difficult for us to distinguish between the element of literary convention and that of actual
emotional stress. When we see Du Bellay, devoured by an impatient longing to see his native France again, use in a sonnet of
Les Regrets, so many different turns of phrase in such an insistent
manner in order to express the subjective length of his sojourn
in Italy, we are entitled to doubt the genuineness of his subjective
attitude to time. Old and new forms of expression occur side by
side in his work:

Depuis que i'ay laissé mon naturel seiour,
   Pour venir ou le Tybre aux flots tortuz ondoye,
   Le ciel a veu trois fois par son oblique voye
   Recommencer son cours la grand'lampe du iour.

Mais i'ay si grand desir de me voir de retour,
   Que ces trois ans me sont plus qu'un siege de Troye,
   Tant me tarde (Morel) que Paris ie revoye,
   Et tant le ciel pour moy fait lentement son tour.

Il fait son tour si lent, et me semble si morne,
   Si morne, et si pesant, que le froid Capricorne
   Ne m'accoursit les iours, ny le Cancre les nuicts.

Voila (mon cher Morel) combien le temps me dure
   Loing de France et de toy, et comment la nature
   Fait toute chose longue aveques mes ennuis.[8]

This poem may be compared with the ballad composed by
Charles d'Orléans during his imprisonment in England, and in
which he expressed his longing: 'En regardant vers le païs de
France,' etc.[9] He was quite devoid of the impatience which
affected the Renaissance poet. For the *rhétoriqueur* time did not

seem to overshadow all. He too had its hopes and longings, but he did not rebel against time and his eyes did not distort temporal relations. We have already said that the Renaissance did not naively follow its subjective feelings, but that it consciously adopted a subjective attitude towards time. External time was its sworn enemy, from which it escaped whenever it could. The practice of beguiling a period of time by informing it with a subjective duration by means of an experience became an art. All manner of turns of phrase now appeared in which this attitude expressed itself. Time was 'cheated' and 'charmed away':

Ronsard:
Afin qu'après ma voix fidelle,
Au soir, à la tarde chandelle,
Les mères, faisant oeuvres maints,
Content tes vertus précieuses
A leurs filles non ocieuses,
Pour tromper le temps et leurs mains.   [II, p. 307]

Des Périers narrated his nouvelles 'pour vous donner moyen de tromper le temps'.[10] This phrase was used of the ability to make a long period of time seem short by means of some activity.

Ronsard:
Pendant que l'heure en donne le loisir,
Avec le vin, l'amour et le plaisir
Charmons le temps, les soucis et la guerre.   [V, p. 356]

Du Bellay:
Ce pendant (mon Ronsard) pour tromper mes ennuys,
Et non pour m'enrichir, ie suivray, si ie puis,
Les plus humbles chansons de ta Muse lassee.   [II, p. 178]

Of all the exponents of the French Renaissance, it was the author of *Gargantua and Pantagruel* who took the greatest liberties with external time. He did not attempt to beguile time, but helped himself to portions of it as he thought fit, extending and contracting it as he pleased. When Rabelais said

Jamais je ne m'assubjectis à heures: les heures sont faictes pour l'homme, et non l'homme pour les heures. Pourtant je fais des miennes à guise d'estrivieres, je les accourcis ou allonge, quand bon me semble[11]

it was not a matter of subjective feelings and impressions, but of an actual distortion and bending of external time. If this attitude is to be understood, it must be realized that the Renaissance was the heir to a positivistic, calculating and date-conscious epoch. The Renaissance, in order to show its true character, had to rid itself of this legacy, which it felt as a burden. It was found that the human heart beat to another rhythm than that of the clock. The world was now a dichotomy, consisting, on the one hand, of an objective and fixed order of things and, on the other, of a complex of human feelings, joys, needs and impulses. The former was rejected in favour of the latter. The manifestations of the new sense of time betokened the awareness of a youthful vigour. Subjectivism was no sign of weakness. Although the attitude of the Renaissance towards time was clearly distinguished from the ancient French conception of it, it must nevertheless be borne in mind that the latter evidenced itself in the meaning of words, whereas the former manifested itself in the content of poetic expression.

The uncertain future

The human need to foresee the future may be fulfilled in two ways: by means of experience, the rational extension of the present into the future and by factual constructiveness on the one hand, and by means of superstitious beliefs on the other. We have seen that the first method of shaping the future was characteristic of the later Middle Ages. As humanity became ever more preoccupied with the future, however, the vogue for calendar-making and astrological nonsense increased, despite enhanced insight into the laws of cause and effect and the nature of the course of things. Although both rational and irrational methods were utilized, even in the age of the Renaissance, in the attempt to foresee what was to come, a reactionary disinclination to pry into the future now made itself felt, based partly on a new awareness of the limits of human knowledge and partly on a newly awakened outlook on life. Sometimes it was philosophy, sometimes religiosity and sometimes temperament that was at the root of this unwillingness to contemplate the future. Numerous sixteenth-century writings, both in poetry and prose, bear witness to the multiplicity of psychological sources which fed this unwillingness. With Du Bartas, it took a religious form: calendar-making was best left to God, who alone knew the hour the clock would strike. Man proposed and God disposed.

Cest luy que tient en main de l'horloge le poids,
Qui tient le Kalendrier, où ce iour, & ce mois
Sont peints en lettre rouge: & qui courans grand'erre
Se feront plustost voir, que preuoir à la terre.

[*La Sepmaine*, I, vv. 381–4[12]]

Bonaventure des Périers gave his satirical poem *Prognostica-tion des Prognostications*[13] a religious and ethical basis when he placed the presumptuous short-sightedness of man beneath divine understanding. The desire to explore the future was heathenish. The *Pantagrueline Prognostication*, on the other hand, was quite devoid of this religious motivation. Rabelais led the fight in the name of nature and common sense. 'Les nonnains à peine concepvront sans operation virile.'[14] In this way he expressed his attitude to the future and advocated that predic-tions should be confined to that sphere in which they were alone infallible: that of scientific fact. The poets of the French Renais-sance did not counter astrology with facts; it was not attacked on scientific grounds; nor was there any talk of legitimate and illicit modes of thought. No objective weapons were brought to bear against the more or less official attitude to the future which then prevailed. It was, however, disliked, because it did not conform to the inner nature of the new humanity, nor to the ideal of life then being popularized. Prognosis was made to appear ridiculous. The gesture of unconcern for the future did not always spring from the same mental attitude; it might be a sign either of strength or of weakness, motivated either by trust in one's own good fortune or by fear. The contrary attitude, that of preoccupation with the future, was informed by the same ambiguity. In the century of Ronsard and Rabelais this turning away from the future resulted from an awareness of one's own strength. One did not behave like the ostrich, for it was less a question of shutting oneself off from the future than of discredit-ing the exaggerated cult of it. There was thus a strong element of the heroic pose in the admonitions of the poets to enjoy the

present; the essence of this pose had its counterpart in the later
Romantics' spineless shirking of the future. Long-term calculations and predictions were contrary to the youthful view of time
of the Renaissance. The future was something irrational, uncertain and intangible, outside the grasp of man. Not only was the
remote future unknowable, but the time immediately impending
was also obscure and contained all manner of possibilities.

Du Bellay:
Va devant à la vigne apprester la salade:
Que sçait-on qui demain sera mort ou malade?
Celuy vit seulement, qui vit aujourdhuy.   [II, p. 194]

Montchrestien:
Qui pourroit du matin juger la fin du jour?
                         [*La Reine d'Ecosse*, II, chorus[15]]

Every civilization tends to divest the future of its element of
novel experience, to make it conceptually understandable, to
weigh up its possibilities in advance and reduce its unpredictability to a minimum. Civilization tames the future like a wild
animal, neutralizes its terrors and guards against its treachery.
Such a view of the future is unpoetic. An attitude to the future
typical of sixteenth-century humanity was to leave the future
'in the air', and passively to commit oneself to it.

This course found its most eloquent advocate in Montaigne.

Elles me seront à l'adventure connues un jour.   [*Essais*, II, 10]

Montaigne was fond of using the future tense in speaking of
uncertainties and things independent of ourselves.

Qui me surprendra d'ignorance, il ne fera rien contre moy. [II, 10]

Quand quelqu'un voudra maintenir qu'il vaut mieux.   [II, 21]

Cherchez qui l'effectuera.   [II, 29]

The irrational and the doubtful appeared to him essentially as
future contingencies. The phrase *à l'adventure*, which he often
used in place of *peut-être*, frequently pointed to the future;

une ame courageusement vitieuse se peut à l'adventure garnir de
securité.   [III, 2]

Of all the Frenchmen of his century he was the one who dealt
most earnestly with the question of man's relationship to the

future. He defined the consciousness of the future of the inexperienced man, who saw it as something essentially different from what would actually come to pass.

The idea of the future and the idea of death were closely associated in his mind. 'What is my future?' was synonymous with the question 'How long have I still to live?' This was not the consequence of an exaggerated fear of death, but of the awareness that the future, even the immediate future, bore within itself an infinite number of possibilities. In writing of himself, he described how, one day when he was a mile from home, in good health and in a cheerful frame of mind, he rapidly wrote down an instruction which was not to be carried out until after his death, because he was not sure of returning home alive. Anything might happen at any moment. This feeling of insecurity did not spring only from the course of the outside world, and from a proneness to any blows fate might strike, but from his confessed state of ignorance both with regard to the changed world he had discovered and to his own self. For him, the future started immediately from the present moment. He was familiar with the irrational nature of time and the helplessness of man with regard to the future. The latter was not a territory that might be surveyed, but something that did not exist. His self-portrait was bounded by the present:

> Je ne peints pas l'estre, je peints le passage, non un passage d'aage en autre, ou, comme dict le peuple, de sept en sept ans, mais de jour en jour, de minute en minute: il faut accommoder mon histoire à l'heure; je pourray tantost changer, non de fortune seulement, mais aussi d'intention. [III, 2]

Thus human knowledge itself contributed to ignorance of the future. Disregard of the present and neglect of current circumstances arising from a too eager concern for the future was a mental attitude divorced from the realities of life.

> La vie est un mouvement inegal, irregulier et multiforme. [III, 3]

We barely understand what is; how, then, can we presume to a knowledge of what is remote in time? Those who, like the milkmaid in a story by Des Périers[16] and the generals of Picrochole,[17] saw the future more clearly than the present were mocked. It is common practice with us today to compare the value of the past, of the present and of the future with one another when we say that insight into the present is more important and valuable than insight into the future, or contrariwise. It is our opinion that the turn of mind which weighs today and tomorrow against each another, and indicates the alternatives of deciding for one or the other, originated in the later Gothic era, which in its inner nature was more concerned with the future than with the present. The Renaissance set this assessment against its own. What is exists by virtue of its presentness and not by virtue of anticipation. One must not be unduly preoccupied with the outcome of an undertaking when embarking on it. When the Cardinal of Ferrara excused himself to Francis I for not having instructed Benvenuto Cellini to make the salt-cellar, on the grounds that the completion of the work was unforeseeable, the king replied: 'Che chi cercassi cosi sottilmente la fine dell' opere, non ne comincerebbe mai nessuna.'[18] Belleau said the same thing in different words:

Qui trop songneux regarde au vent, iamais semence
Ne fera qui profite: et qui sous l'inconstance
De l'air se veut regler, espiant les saisons
Iamais ne iouyra de fertiles moissons.
[III, p. 196, *Discours de la Vanité*]

Every undertaking extending into the future was a wager, and the Renaissance, confident that fortune would not desert it, accepted the wager. In that century, when what was present and seemingly familiar and simple became mysterious and complex for those who had deserted medieval modes of thought, when Montaigne adumbrated, not miracles, but the marvel of heredity, when the world had expanded and a welter of questions cropped up which the Middle Ages had not posed, the gaze of humanity was involuntarily averted from the future, because the present was a living entity which clamoured for attention. The strongest voice was again that of Montaigne, when he inveighed against

la forcenée curiosité de notre nature, s'amusant à préoccuper les choses futures, comme si elle n'avoit pas assez affaire à diriger les presentes.   [I, 11]

Although the Renaissance had discovered the individual, and although the individual had gained an enhanced awareness of his spiritual and mental qualities, he had nevertheless become more humble with regard to the powers directing human life. A profound realization, springing from immediate intuition, told him that his destiny was not the effect of externally visible and calculable forces. Individualism, as it awakened, strained at the confines of the human, and experienced them in a more human way than did the Middle Ages. In the Middle Ages nothing was said of the inscrutability of the future because this inscrutability then attracted no attention. In the sixteenth century the limitation of the human horizon to the present lost its individual nature and became a general world-view and conception of life, finding its expression in the injunction *carpe diem*. Present enjoyment, and not hope, was the source of felicity.

R. Belleau:
Desirer est un mal, qui vain nous ensorcelle:
C'est heur que de iouir, et non pas d'esperer:
Embrasser l'incertain, et tousiours desirer
Est une passion qui nous met en cervelle.   [I, p. 106, *Le Desir*]

Que de beaux iours coulent soudain
Comme la fleur, et qu'il faut prendre
Le plaisir sans le surattendre
Ny le remettre au lendemain.   [I, p. 153, *D'un bouquet*]

This *carpe diem* sprang from a new insight into the nature of time. The phrase implied that time could be turned to account only by committing oneself to it, and by abandoning the objective order of time as far as possible. True duration existed when hours and minutes went uncounted. This cult was diametrically opposed to the one which concentrated its attention on the idea of having, gaining and wasting time. He who bade the moment 'Tarry awhile, for thou art so fair' had the most profound experience of time.

Whereas, on the one hand, the anxious and calculating exploration of the future was so impulsively reacted against, on the other a novel human attitude to the future was represented in certain domains of poetry. This consisted in the anticipation of coming events by such phrases as 'I already see . . .' There were few such turns in medieval literature, although they expressed a basic human impulse—the anticipation of what was longed for or

feared—by means of an overwhelming feeling, impatience or
fear ('He already saw himself in the hands of the enemy'). The
sixteenth-century poets made of this anticipation, formerly
incidentally expressed and affectively conditioned, a rhetorical
image and imparted to it the ceremoniousness of prophecy. In
France this insight into the future, arising from poetic vision,
found its chief exponent in Ronsard:

> Je les enten deja tonner [tes vers]
> Dedans le palais ce me semble,
> Et voi nos Poëtes ensemble
> D'un tel murmure s'étonner.
>   J'entrevoi deja la lueur
> Des bien estincellantes armes,
> Chasser en fuite les gensdarmes
> Et les chevaus pleins de sueur.   [A René Macé[19]]

The individual, who otherwise so consciously desires to live in
the present, is here represented as one who forgets the present
moment in a vision of impending events. While calendar-making
was condemned, one derived pleasure from playing the part of one
whose eyes and ears actively took possession of the future. The
forecasting of the future, which in life was discredited, had
retreated into literary fiction and produced rhetorical blooms in
this sphere. In turning to antique models the Renaissance discovered the aesthetic value of prophecy. At no stage in the
history of European literature did soothsaying find such extensive expression in poetry. The poet used it in order to represent
current circumstances as having been long since foretold by a
seer. Often, prophecy was a means of paying homage to a living
ruler and patron, projected into the future as the scion of an
extinct lineage.[20] Ronsard put the following prophecy on the lips
of the Fates, gathered round the cradle of Henry III:

> Nous voyons la Charente, et les bords d'alentour
> Desjà rougir de sang, et l'air de Montcontour
> S'infecter de corps morts, et ses plaines semées
> D'os, porter à regret les mutines armées.
>   Desjà nous te voyons au milieu des travaux
> Renverser à tes pieds chevaliers et chevaux,
> Et pendre sur ton front pour eternelle gloire
> L'honneur et le bon-heur, la force et la victoire. [*Les Parques*[21]]

The function performed by time in such prophecies lies in throwing into relief that which is historically significant. What can be foretold has a certain sublimity. The feeling for time that begets prophecy is no everyday form of thought, but belongs to the realm of free imagination. Towards the end of the sixteenth century the more gloomy element became uppermost in poetic soothsaying. Cassandra held sway:

> J'entens desja la voix du juge inexorable,
> Je voy desja l'apprest du tourment perdurable
> Qu'ont pour les malheureux les demons estably.
> [Racan, *Les Bergeries*, V, 2[22]]

The *motif* of the vision of evil and death was borrowed by the Baroque drama from Seneca:

> Agamemnon ie voy le poignard dans le flanc,
> Contre terre estendu se souiller de son sang,
> Se mouuuoir, se debate, ainsi qu'un boeuf qu'on tue,
> Apres le coup mortel s'efforce, s'éurtue,
> Se tourne et se retourne, et par ce vain effort
> Cuide se garantir de la presente mort.
> Puis ie voy la fureur du parricide Oreste,
> Comme sa mere il tue, et le fils de Thyeste,
> Et comme transporté d'amour hymenean,
> Pyrrhe il va massacrant, le meurtrier de Priam.
> [R. Garnier, *La Troade*, vv. 335–44[23]]

Side by side with such prophecies appeared awesome expressions of anticipation of impending evil:

> Ie voy ia de Charon la teste cheuelue
> Et les larues d'Enfer, i'entens l'horrible voix
> Du chien Tartarean hurlant à trois abois.
> [Garnier, *Antigone*, vv. 1137–19[24]]

> Dans mon esprit troublé passent incessamment
> Mille horribles pensers iuges de mon tourment;
> Or' ie me voy trahi de mes propres gensdarmes.
> Or' ie me sens tomber au plus fort des alarmes.
> [Montchrestien, *David*, II[25]]

In such utterances, often decked in mythological garb, the future casts a shadow over the present. The ideal of untroubled involvement in the moment is here forgotten.

The art of life

The more the outlook of medieval humanity was turned towards this life, the greater became the freedom of the individual to mould it according to his wishes, both in thought and in reality. A longing for the beautiful life was a spiritual necessity for the fourteenth and fifteenth centuries, which had become subject to a dispirited frame of mind.[26] One of the basic needs of the individual who consciously strove to brighten his existence was to have time at his disposal. It was a matter of making of this time, which slipped by amorphously and regularly, and which was neither beautiful nor ugly in itself, an instrument of pleasure by using it (in the aesthetic sense) as purposefully as possible. It was a *sine qua non* of this mode of life to establish a suitable temporal relationship between one's activities, and to strike a temporal balance between work and rest, and between matters grave and gay. This idea of an appropriate alternation of activity was an indispensable part of the dream of the beautiful life.

This free deployment of time, which made of life an artistic creation, was of the utmost interest to the humanity of the Renaissance. Even in earlier centuries, however, suggestions for the deployment of time had been made. Philippe de Novare proposed the following as an ideal daily routine. As soon as one awoke after midnight, one crossed one's chest and said a prayer. After rising early, one attended divine service, followed by the giving of alms. On returning home, one attended to one's toilet, etc., after which one turned immediately to matters of business. These should be disposed of as soon as possible, so that the day's work was completed by noon. The mid-day meal was followed by an hour's rest. The afternoon should be given up to recreation and innocent pleasures, and the evening spent in company. The night-time was for repose.[27] This was the method of dividing the day of a Christian and sage, in which moral and aesthetic considerations, duty and recreation were equally balanced. It was a mean between two extremes: the monastic way of life on the one hand, in which the division of the day was dictated exclusively by the service of God and those nearest to one; and the Rabelaisian mode of deploying time on the other, the principle of which was summed up in the phrase *fais ce que vouldras*. Philippe de Novare observed moral rather than aesthetic criteria in his method of dividing the day, because the agreeable afternoon recreation of which he spoke had as yet no great intrinsic value, for recreation and pleasure fulfilled a definite moral purpose and were of the nature of a duty:

> Et après se doit on delitier en aucune chose, por avoir remede et repos en son cuer, et aaisier soi sanz pechié et sanz honte et estre la vesprée aucontre la gent por veoir et aprendre et dire et faire aucun bien honoréement, selonc ce que Dieus li a doné de sa grace.[28]

The distinction between recreation with a moral purpose, desired and permitted by God, and that indulged in purely for the sake of disporting oneself, was a medieval one. All pleasures shared in company were subject to religious considerations. One should take one's enjoyment 'ad necessitatem, non ad voluptatem'.[29] But the way in which the youths and young women of the *Decameron* deployed their time indicated an altogether aesthetic evaluation of leisure. The Italians adopted this outlook before it became current in France. The most striking practice of fifteenth-century humanity in France was the secularization of the method of apportioning time, as a step towards an exclusively aesthetic mode of deploying it. The useful life still seemed more desirable than the beautiful. Historians now showed a keen interest in the manner in which princes employed their time. A description of the daily activities of a person was a stroke added to his portrait as an individual. Chastellain reported of Charles VII that he

> mettoit jours à heures de besongner à toutes conditions d'hommes, lesquelles infailliblement vouloit estre observées, et besongnoit de personne à personne distinctement, à chascun: une heure avecques clercs, une autre avec gens mechaniques, armuriers, voletiers, bombardiers et semblables gens; avoit souvenance de leurs cas et de leur jour estably: nul ne les osoit prévenir.[30]

This was a thoroughly practical way of using time, in keeping with the natures of Charles VII and Louis XI.

At the court of Burgundy another spirit held sway: the same historian related of Charles the Bold that he

> tournoit toutes ses manières et ses mœurs à sens une part du jour, et avecques jeux et ris meslés, se delitoit en beau parler et en amonester ses nobles à vertu, comme un orateur. Et en cestuy

regart, plusieurs fois, s'est trouvé assis en un hautdos paré et ses nobles devant luy, là où il leur fit diverses remontrances selon les divers temps et causes.[31]

Three times a week—on Mondays, Wednesdays and Fridays— Charles the Bold kept court, with great outward display.[32] From this instance also, it can be seen that the deployment of time was for him less a matter of utility than of a need to inform time. The method by which Charles the Bold shaped time was aimed more at the exalted than at the beautiful, and at what was immutable, imposing and majestic rather than pleasant or agreeable. A method of dividing time which goes beyond utilitarian considerations is either beautiful or sublime.

By a 'beautiful' mode of apportioning time we understand one which strives for the pleasant, and which accepts those divisions of time dictated by natural laws, such as day and night, which aims at alleviating physical and mental fatigue, and which sets itself to supply the demand for a more agreeable way of life in a versatile and adaptable manner. Humanity and its desire for happiness, ease, repose and pleasure appear to those who hold this view of time more important than a rigid temporal order to which man is subject.

Adherence to a set temporal scheme is, however, exalted if it serves an aesthetic purpose. In this matter the individual became aware of his proper place. He had no rights. An abstract ideal of the division of time held sway. Such things as ceremonies, processions, celebrations, visits and theatrical performances, the time of which was precisely fixed, and even a temporally regulated mode of life considered as a whole, wherein the individual, beset by fatigue and changes of mood, was relegated to the background, were exalted also. One might in this connection adduce the case of a public entertainment, where the prime consideration was not individual enjoyment but rather that things should go off in an orderly manner and according to plan; in this its exalted quality resided. Thus a method of deploying time that was more sublime than beautiful was characteristic of the Middle Ages, because in that era the individual effaced himself, subordinating personal considerations to the ideals of Christianity. The inflexible sublimity with which the medieval view of life was informed was typified by the notion of the four ages of life as expounded in the book of Philippe de Novare. In it the divisions of life were set out in juxtaposition, but with no form of link or transition between them. They were not stages in a continuous process, but self-contained components of a vast edifice.

This division was not a conceit springing from a realization of the empirical nature of the course of life. It was nothing less than a challenge: this was the pattern to be followed by an exemplary life. The right conduct of life was no problem for the Middle Ages. It did not—indeed it could not—become problematic until such time as humanity became aware of the freedom it was vouchsafed in life, and until one realized that one should—nay, must—decide oneself the course of one's life. Only then did the norm recede, its place being taken by an indefinite number of different possibilities for shaping one's life. Everyone lived his particular existence, each had to fill up and divide his time according to his own judgment. In order to do this he had to know himself.

The question of how best to live and deploy one's time was now for the first time posed free from metaphysical complications. It is not surprising that the phrase 'way of life' was added to the language. In Montaigne we read:

On leur va ordonnant une non seulement nouvelle, mais contra-dictoire [var. *contraire*] forme de vie. [III, 13]

Ma forme de vie est pareille en maladie comme en santé. [III, 13]

The art of living was a new obligation which now imposed itself; indeed, it had become the chief concern of mankind. Over and above all demands arising from the particular circumstances, needs and natures of individuals, no matter how great and important these considerations might be, Montaigne placed the general duty of coming to terms with time, with one's lifetime, and of regarding life itself as a task that must be performed well.

Avez-vous sceu prendre du repos? vous avez plus faict que celuy qui a pris des empires et des villes. Le glorieux chef-d'œuvre de l'homme, c'est vivre à propos; toutes autres choses, regner, thesau-riser, bastir, n'en sont qu'appendicules ou adminicules pour le plus. [Var. 1595]

Order must be imparted to life, but, on the other hand, it must not be forced into the inflexible mould prescribed by the medieval mentality. In either case the individual would be unfree, for he would be living to order rather than living his own life. He would be the object of life or of an external scheme of things. Now, however, man became aware that time and life were in his own hands, that they were empty and shapeless in themselves, and that he had to fill and inform them according to his personal sense of liberty and responsibility. The exercise of this foremost and most exalted art did not call for any special abilities or knowledge: indeed, these might stand as obstacles in his path. The particular gifts, intentions and activities of an individual might injure his existence as a human being; they might render him excessively pedantic or dictatorial, and might cause him to neglect the elementary demands of human life. He must see his life-goal, not in what he did, but in himself. If his thoughts and efforts were directed towards something entirely outside himself—whether it was a war to be won, a book to be written or a profession to be exercised—he would thereby violate and degrade his time and life by devoting them to something external and alien. In the eyes of the Renaissance, moreover, time did not exist to enable a duty or task to be fulfilled, but for man to use as he thought fit. One's life was an end in itself. One must be aware at every moment that one was living through time, that one held it in one's hands and that one was able to grasp it firmly or let it escape.

> J'ay un dictionnaire tout à part moy; je passe le temps, quand il est mauvais et incommode; quand il est bon, je ne le veux passer, je le gouste, je m'y arreste: il faut courir le mauvais et se rasseoir au bon, [III, 13]

said the author of the *Essais*. Time was what one was able to make of it. The idea that time was a substance which might be shaped by human will showed itself in the language in the form of new words and phrases. In the fifteenth century, *passer le temps* and *passetemps* appeared in association:

> Chastellain:
> En tels festiemens et bonnes chières fit-on passer au dauphin son temps en Bruges à sa première venue. [III, p. 310]

> Là chassoient et passoient temps. [III, p. 324]

> Roi René:
> Ainsi le temps gentement passeront.
> [II, p. 111, *Le Pas de la Bergiere*]

Cent Nouvelles Nouvelles:
Quand la brigade fut trèsbien repue, la cloche sonna XII heures,
dont ilz se donnèrent grans merveilles, tant plaisamment s'estoit
le temps passé à ce souper.   [7]

Passa le temps comme il souloit.   [10]

Qui miserablement son temps passoit.   [11]

Ce vaillant homme va passer temps.   [12]

Gringoire:
Quelque part va le temps passer.33

Cent Nouvelles Nouvelles:
Après ce bel passetemps.   [9]

Les plaisans passetemps qu'elle souloit avoir.   [22]

En ce trèsglorieux estat et joyeaux passetemps se passèrent pluseurs
jours.   [13]

It is evident from these examples that here was a new means
of expressing the idea of recreation and of pleasure shared in
company. Words formerly used in this sense were *esbanoier*,
s'esbatre, *se solacier*, *se deduire*, *se joer*, *se deliter*, *renvoisier*,
dosnoier, among others. The fact that 'pastime' was used in their
stead in the fifteenth century shows that social pleasures were
seen as a method of filling a particular span of time. The new
term sprang from an awareness of the psychological fact that
time, when devoted to pleasure, passes most rapidly and is a
fertile source of impressions. The word reminds us of an observation earlier made: time—if not an entire lifetime, then at least a
certain period of it—appeared as something to be worked on.
The individual felt his attitude to it becoming increasingly
active. But whereas *passer temps* was still somewhat a matter of
convention in the fifteenth century, Montaigne raised the concept of passing time to an art, one which he practised with
mastery. Time was something he valued more greatly than did
the dying Gothic age.

Cette fraze ordinaire de Passe-temps et de Passer le temps re-
presente l'usage de ces prudentes gens qui ne pensent point avoir
meilleur compte de leur vie que de la couler et eschapper, de la
passer, gauchir, et autant qu'il est en eux, ignorer et fuir, comme
chose de qualité ennuyeuse et desdaignable.   [III, 13]

If 'passing' time meant no more than 'killing' it, the humanity of the Renaissance, pervaded by a new sense of time, necessarily felt impelled to revolt against the word and the thing it stood for. Nothing was effected by merely passing time. It must be animated, filled, savoured and turned to account, not in extension but in depth.

Each of the great minds of the age saw time as a task to be performed. It must be recognized as an organism and treated appropriately. Just as organic entities existed in extension in the world, so it was possible to speak of temporal organisms. A day, a year, a human life, the birth and death of a culture were such organisms. Such historical and natural organisms could, with relative ease, be apprehended as wholes unified by their element of inevitability and by the coherence of their constituent parts. They were primarily objects of perception. The idea of an organism might, however, also appear as a challenge and a goal. An activity, visualized as a whole contained by time, might be mapped out and made a reality. Its elements belonged together and must not be separated. Enemies of temporal organism were distracting interruptions, dissipation of effort, and the forcing of some process into a temporal framework inappropriate to it.

The Renaissance, with its new awareness of temporal organism, resented the regulation of life by the clock as an arbitrary and odious fragmentation of organic forms. This fragmentation prompted Rabelais' revolt against the clock. He did not accept the normal temporal order as it existed previous to himself as an established institution of life, but viewed it as a canvas on which to paint a fancy-free picture of new, as yet unrealized possibilities. His art of living strove for complete liberation from the trammels of time. Clocks and calendars were not the affair of free persons. The ideal was to enjoy time as an experience. The inmates of the abbey of Thélème were blissfully ignorant of the rigid temporal order. Fully conscious of his new view of time, Rabelais proclaimed his art of living.

Et, parce que es religion de ce monde, tout est compassé, limité, et reiglé par heures, fut decreté que là ne seroit horologe, ny quadrant aulcun. Mais, selon les occasions et opportunitez, seroient toutes les œuvres dispensées. 'Car, disoit Gargantua, la plus vraie perte du temps qu'il sceust estoit de compter les heures,—quel bien en vient il?—et la plus grande resverie du monde estoit soy gouverner au son d'une cloche, et non au dicté de bon sens et entendement.'

[*Garg.*, I, 52]

The ideas of 'freedom' and 'time' were closely associated in Rabelais' mind; that is to say, he understood human freedom as freedom from the tyranny of time. His celebrated injunction 'Do what you will' was meant in a predominantly temporal sense:

Se levoient du lict quand bon leur sembloit, beuvoient, mangeoient, travailloient, dormoient, quand le desir leur venoit.   [I, 57]

Likewise:

Si quelqu'un ou quelqu'une disoit: 'Beuvons', tous beuvoient. Si disoit: 'Allons à l'esbat es champs', tous y alloient.   [I, 57]

Freedom thus resided largely in being able to act when one chose. The idea underlying the above excerpts was a total recasting, born of an overweening optimism, of the relationship of man to time. The choice of the right moment to act was trustingly left to an individual who had attained his majority. Living by the clock was felt as a distortion of the natural and beautiful way of life. Time must not be allowed to become a tyrant. The natural order of things was better than any temporal constraint. Our lives should be shaped by our personal preferences, our moods and our discretion. If we follow our own devices in fixing the time of our activities, then our decision is one consonant with our liberty. The following epigram of Marot, which reflected a general desire, was in keeping with the spirit of Rabelais:

Du temps present à plaisir disposer.   [III, p. 49]

For Rabelais there was no antithesis between the individual and the community. What the individual did with his time without external constraint, contributed to the benefit of all. Rabelais had no fear of the consequences of individualism; he was untroubled by the thought that the community would disintegrate outwardly and inwardly as a result of individual waywardness and the lack of temporal co-ordination in communal life. The inner clocks of individuals were seldom in agreement.

As though in self-correction, Rabelais set this limitless freedom of the individual in the matter of time against an image of the most extreme 'time snobbery'. The schoolboy Gargantua, whose education we witness, was subjected to a sort of Taylorian system so that he might not waste a single minute: he rose at four in the morning, whereupon he underwent a rub-down and listened to a reading from the Bible at the same time. The passage read was recapitulated 'es lieux secrets'. While he dressed, combed and perfumed himself he repeated what he had learnt on the previous day. We observe this same simultaneity in attending to scholastic and other matters during the mid-day meal and at other times of the day.[34] Whereas Rabelais banished clocks from the abbey of Thélème, here he fills every minute. In the abbey, time was an experience; for Gargantua, it was a quantity. In the one case it meant freedom, in the other servitude. When he attacked the pedantic and inflexible temporal order of the monasteries Rabelais ran the risk of over-stating the case for emancipation from the exigencies of time. It might seem, that in trying to escape from the Scylla of regulation by the clock, he fell victim to the Charybdis of unorganized, amorphous, lost and wasted time. But he attacked the latter monster with the same panache and extravagant energy as the former.

It would be a mistake to interpret one or other of these two opposing attitudes as reflecting, unequivocally and exclusively, Rabelais' own, and to identify oneself with the one in order to refute the other. They both represent two different but equally noteworthy facets of his nature. One shows him as the champion of the Renaissance, rebelling against the monastic tyranny of the clock because it confined the individual in time just as the monastery walls confined him in space. The other shows him as a humanist and cultural fanatic who did not regret a single minute devoted to civilizing pursuits. The individual whose keenest pleasure springs from the acquisition and exercise of culture, and the humanist, who evaluates all human possessions in their relation to culture, will see the need of having time at their disposal as an axiomatic prerequisite of their intellectual activities. The time at the disposal of medieval humanity formed, at least as far as moral obligations were concerned, an integral part of the general order of the world, which existed as a stable institution.

However, the relationship of the individual's time to the ordering of communal life showed a reversal of this position, in that the latter was conditioned by the former. Rabelais' impetuosity drove him to adopt an attitude to time as ruthless in his personal life as in his literary creations. But what would have been for others servitude to time and slavish conformity to the clock was in his case a manifestation of strength. In truth, he did what he wished with time. In his hands the clock was no longer an instrument of temporal frugality. He was well able to discard it, for he was independent of measured time, even if he caused Pantagruel to use carrier pigeons 'pour racheter et gaigner temps'.[35] His thought embraced two diametrically opposing attitudes to time: total disregard of it at one extreme and total subservience to it at the other. Sometimes the living human being, 'el hombre de carne y huesos', was uppermost in his sense of values; sometimes he saw only the abstract ideal: the fullest use of all means in one's efforts to acquire intellectual riches.

An innovation introduced by the Renaissance—and, indeed, one that was peculiar to it—was the anthropocentric view of time. Man was now an end in himself. He did not exist for the sake of institutions; they existed for his sake. They had no *raison d'être* in themselves. Objective time now had to adapt itself to the purposes of humanity, whose natural needs and empirical characteristics now stood out more clearly. Humanity formed an element of nature, communing in vast rhythmic harmony with the whole of creation. A flexible way of life, adapting itself to circumstances and personal considerations, took the place of the rigid medieval attitude to time. Ambroise Paré represented flexibility and adaptability in the temporal shaping of existence as a natural necessity:

Le temps d'avoir assez dormi, se cognoist à la parfaite concoction des viandes, et non par certaines heures determinées: car aucuns cuisent plus tost, les autres plus tard, combien que le plus souvent la concoction se fait en sept ou huit heures.

[I, p. 72 (*Du dormir et veiller*)]

In this century one did not need to be a doctor to advocate the natural apportioning of time. The new art of living did not emanate from specialized medical knowledge but from a belief that mind and nature had to be attuned to each other, and that to have time at one's disposal according to the requirements of nature would benefit one's higher mental faculties. The dictum of the society queen Gostanza on the proper use of time, incorporated by Firenzuola in the preface to his short stories, forms the classic expression of the necessity of attuning one's mind with nature:

> E poichè noi semo sei, e vogliamo stare qua sei dì, io vi voglio dividere il giorno in modo, che ogni nostra opera proceda per sei. E perciocchè la mattina lo ingegno suole esser più svegliato che di niuno altro tempo, e' sarà bene che andandoci a spasso or su questo monticello e or su quell'altro, noi ragioniamo di qualche cosa, che sappia più delle scuole de' filosofi che de' piaceri che ne sogliono apportar le ville.[36]

It seems that the Italians tackled the question of dividing time with more professional skill and more as a matter of routine than did the French. The words spoken by Callimaco in Machiavelli's comedy *Mandragola* indicated an assured and purposeful weighing up of the question of suitably apportioning time to one's various activities:

> Havendo compartito el tempo, parte alli studii, parte a'piaceri, e parte alle faccende; e in modo mi travagliavo in ciascuna di queste cose, che l'una non mi impediva la via dell'altra.   [I, 1[37]]

This sense of proportion in the division of time was in keeping with the spiritual balance of the Italian of the Renaissance, who was inclined to dilettantism rather than to pedantry. We do not find this architectonic method of shaping one's way of life formulated by any Frenchman of the sixteenth century. The closest approach to the routine deployment of time of the Italians was to be made by Montaigne.

The phrases 'suitable time' and 'right moment', as used in the sixteenth century, often carried a meaning different from that which they bore in the later Middle Ages. A suitable time was not one which offered a favourable combination of external circumstances, but was a point in time most in harmony with the natural order of things and the prescriptions of the natural way of life. Paré said that 'le temps le plus commode de dormir est la nuict'.[38] Medieval humanity slept at night, because God had intended it and ordained it for rest. Only the evil-doer kept awake. The view now prevailed that night was the natural time for sleep; it had its place and function in the natural order of things, which was now acknowledged. The word *commode*, when used with reference to time, meant 'natural', 'corresponding to human needs'.

> Et leur dictes qu'ilz trouvent maniere de departir leur temps le plus commodement pour vous et pour eux qu'il sera possible.
>
> > [Des Périers, II, p. 37 (Nov. 6)]
>
> Et depuis continuèrent leurs affaires ensemble à toutes les heures que le clerc trouvoit sa commodité.   [*Ibid.*, II, p. 45 (Nov. 8)]

This way of thinking reflected an endeavour to shape time to fit one's varying individual needs.

The greatest exponent of this art was Montaigne. What he said of Socrates' way of shaping life was his own ideal of life:

> Il dira qu'il sçait conduire l'humaine vie conformément à sa naturelle condition.   [III, 2]

He was able to give, as it were, individual treatment to the moment, and to shape the possibilities inherent in experience into their most expedient form. He listened to his inner voice, he eavesdropped on his own rhythm of life, in order to select the appropriate time for doing something. Let us hear how he lived with his books:

> Ce sera tantost, fais-je, ou demain, ou quand il me plaira... Ils sont à mon costé pour me donner du plaisir à mon heure... Là, je feuillette à cette heure un livre, à cette heure un autre, sans ordre et sans dessein, à pieces descousues; tantost je resve, tantost j'enregistre et dicte, en me promenant, mes songes que voicy... Je vis du jour à la journée, et parlant en reverence, ne vis que pour moy: mes desseins se terminent là.   [III, 3]

The programme which Rabelais set out in his Utopia here became a reality; albeit not in the form of communal life.

Montaigne divorced a present action from the purpose it might fulfil in the future. The action was sufficient in itself. The tense expectancy intervening between preparations, arrangements and precautions on the one hand, and fulfilment of one's purposes on the other, was reduced to a minimum. The future was, as it were, eliminated from consciousness. It was not required. Montaigne resembled the traveller with no definite idea of the destination he intended to reach. There were so many attractive spots on the way which invited one to linger awhile, and which were not to be regarded simply as points through which one passed *en route* for a destination which must ever be kept in mind. The ordering of life by means of a rigid temporal scheme was contrary to the Renaissance conceptions of time and of the individual. Such an ordering deprived the individual of the capacity to react in an adaptable manner to the ever-changing demands of the moment. This represents a clear and simple definition of an idea which, at the time, existed in thought and language only as a tendency, a desirable state and an ideal of life.

To be sure, Montaigne's trivial preoccupations with the appropriate time for reading were much less weighty and significant than his attitude to time as it affected the important and decisive moments of life. But even these were not the affair of the calendar or of pious symbolic contemplation; they were, rather, voices which claimed the attention of an ear sensitive to the changing melodies of life. Montaigne's thoughts tended to dwell on the period of transition from manhood to old age. One of the most essential and important duties of the art of living was precisely to discern the point at which one entered into old age, and judiciously to adjust the tone of life to one consonant with the course of nature, so that no jarring note might be struck. The decision of Charles V to step from the stage of the world when he realized, in his declining years, that he was no longer able to play an effective part in affairs, was praised by Montaigne as the most laudable action of his life.

The face of time

In the later Middle Ages the generalized idea of time, *le temps*, existed. The use of the definite article with a 'presenting'[39] function denoted that the idea of time had become abstracted. In Old French the word existed only with the meaning of particular sections of time and the particular circumstances associated with them in imagination. Now, however, time gradually began to acquire the form of a philosophical concept, which became a subject of reflection and which was acknowledged as a general condition of human existence. Even in the later Middle Ages, all manner of special qualities were associated with time, but as a comprehensive notion it played only an unimportant part in consciousness. Time was not yet seen as the destiny of the individual, although it is true that in the fourteenth and fifteenth centuries the thought of transitoriness had inspired humanity with dread. It was a way of thought characteristic of that epoch to see and experience the brevity of human life and its dwindling into nothingness, not from the point of view of time, but from that of death. The idea of time had as yet no association or contact with the concepts about which philosophical thought revolved; indeed, the word *temps* became divorced from the idea of 'life' in Middle French: the equation 'time = life', current in the age of Old French, began to disappear.

If the imagination of the later Middle Ages pictured death so vividly and time so dimly, this was doubtless because death is a more 'immediate' notion than time, that is to say, it presents itself to consciousness more readily than the latter. It is a highly pictorial concept, a unique event, and one which demands to be called by name. In the *Chanson de Roland*, *la mort* was the only abstract concept to be used with the definite article. 'Death' is a much more vivid word than 'time'. In proverbs it occupied a much more prominent place than the latter. In the language of the Bible and of medieval piousness it was a sort of symbol for any form of transitoriness. Time is a philosophical, attenuated and colourless concept, and was used less in reference to one's own destiny than was the thought of death. Villon, that personal lyricist, had no occasion to employ the word *le temps* in his verse, although when at the height of his powers he did inveigh against transitoriness and the effects of time. It is probable that time impressed itself more strongly on consciousness when thought turned away from the fate of one's physical body and became engaged to a greater extent by the perpetuation of one's personal works and values here on earth. This necessarily caused the image of death to wane; in the light of this reorientated outlook it appeared less as a severance and a gulf, for the idea of posthumous fame, which was immune from death in the physical sense, was added to that of physical life. Thus in the Middle Ages it was the awareness of the desirability of achieving fame which first gave rise to the philosophical concept of time. This turn of thought visualized death in forms other than the physical.

Petrarch called the destruction of one's grave a second death, and the sinking into oblivion of one's own books, and those of others in which one's name was alluded to, a third. It was time, however, that brought mental death. Physical death was nothing, a trifling circumstance relatively to time and its effects. Petrarch, a philosophical poet, spoke less of death than of time. This was a true subject for the humanist, who was mindful not only of time but also of fame. It was, therefore, an ancient and non-Christian way of thinking which induced the Italian poet to become conscious of time and to set a high value upon it. This mode of thought came into its own in the sixteenth century. Time was an important element of Renaissance thought; indeed, it formed a part of its general consciousness. The problem of time took its place amongst the basic considerations which played on human emotions—life, death, God, nature, history and eternity. The practical attitude towards it characteristic of the fourteenth and fifteenth centuries was succeeded in the age of humanism by a philosophical one. Technical considerations— the deployment of time and the means of orientating oneself in it by its spatial projection—gave way to a humanistic view of time. It now belonged to another conceptual sphere than formerly. Its nature was inquired into, an attempt was made to characterize it more individually than before, and there was a realization of the important part it played in human life. The Renaissance approached the concept of time from the most diverse 'angles'; it touched upon it in connection with all matters which engaged its attention.[40] It was endowed with a face and other features by the poetic language of the century. It was represented as something active and human. Phrases such as 'time heals all wounds', 'time brings counsel', 'time overcomes all obstacles' and so on became increasingly numerous during the sixteenth century and took a permanent place in the language.

Ronsard:
Le temps, qui les prez de leurs herbes
Despouille d'an en an, et les champs de leurs gerbes.   [I, p. 184]

Au milieu d'elles [des Parques] estoit
Un cofre, où le Temps mettoit
Les fuzeaux de leurs journées.   [*Odes*,[41] I, 10, epode 19]

Et des chansons qu'a voulu dire
Anacreon desur sa lyre,
Le temps n'efface le renom.   [*Odes*, I, 16]

   le temps, qui peut casser
Le fer et la pierre dure.   [II, p. 227]

Belleau:
Possible le temps fera naistre
Quelque nouvelle occasion.   [III, p. 324]

Du Bellay:
Le Temps, qui tout devore.   [I, p. 154]

Le Temps qui tousjours vire.   [I, p. 156]

Time was represented as assuming the most diverse guises.
Once it was seen as an avenger:

Ronsard:
O de Paphos et de Cypre regente
  Deesse aux noirs soucis,
Plustost encor que le temps sois vengeante
  Mes desdaignez soucis!   [II, p. 214]

And once as a destructive force:

Du Bellay:
Temps injurieux.   [I, p. 98]

L'injure du temps.   [II, p. 398]

Soubz le grand Espace
Du Ciel le Tens passe
Par Course subite:
Theatres, Colosses
En Ruines grosses
Le Tens precipite.   [I, p. 184]

Its destructiveness might appear as that of a scythe:

Ronsard:
Mais allonger son nom, et le rendre aimantin
Contre la faulx du Temps, dependoit du Destin.   [IV, p. 217]

In his *Hymne de l'Eté*[42] Ronsard portrayed time as an enfeebled old man. In a sonnet by the same poet it appeared as a man with winged heels, armed with a knife, and whose hair, in which a bald patch was present, hung down over his eyes. This was an imitation of Posidippos.[43]

The irrational and intangible qualities of time, which now seemed enigmatic and treacherous, found their expression in many formulations used in the sixteenth century to characterize time. Lorenzo the Magnificent said of it that 'Perchè 'l tempo fugge e 'nganna.'[44] Time slipped elusively away, it was capricious and played its little tricks. In the final sonnet of his *Deffence et Illustration de la Langue françoyse*, Du Bellay spoke of its wickedness: 'du Tens la malice'. There was no getting the better of time.

Ronsard:
Le temps leger s'enfuit.   [I, p. 155]

   l'heure s'enfuit
D'un pied leger et diligent.   [II, p. 469]

Le temps s'enfuit, le temps qu'on ne rattrape
Quand une fois des mains il nous eschape.   [V, p. 290]

The properties associated with time were specifically feminine ones. It was gay, fleet-footed and gracious. Its melody was not grave and heavy; it made no dramatic pauses; the image of time was without any markedly tragic quality. The hour, that relentless tyrant of the medieval and monastic way of life, was portrayed by Belleau, with his taste for antiquity, as something much more amiable, with its qualities of fleetingness, grace and humanness. The passing of the hours was a brief melody, devoid of any heroic quality. The poet was aware, nevertheless, that the inexorable effects of time were at work behind this agreeable façade. Thus he said to the Hour:

Toute la force et la grace
Du ciel se remire en toy,
Et la violante audace
Du temps ne gist qu'en ta foy.[45]

The Renaissance poet Belleau, who was susceptible to impressions rather than to facts, and to the melody of events rather than to their effects, was more conscious of the varied and gay succession of the hours than of their power of bringing about change.

The realization that time carried all away was no doubt peculiar to the Renaissance, but it did not exert a dispiriting influence. The Renaissance gave forth only a subdued and dolorous undertone in its knowledge of man's impotence with regard to time. The reflection that time, once flown, would never return likewise drew forth another note, akin to the former. This imparted an enhanced value to time, albeit not the value it possessed in the eyes of the bourgeois later Middle Ages, but one arising from its uniqueness as an experience, more particularly in youth. The irrevocableness of time was a more serious matter than its loss. Time as a quantity could be replaced, but time as an experience could not, nor would it ever return. The poets were fond of dwelling on the theme of the uniqueness of experience.

Ronsard:
Dieux, vous estes cruels, jaloux de nostre temps!
Des dames sans retour s'envole le printemps,
Aux serpens tous les ans vous ostez la vieillesse.   [I, p. 341]

Nos ans sans retourner s'en-volent comme un trait,
Et ne nous laissent rien sinon que le regret
Qui nous ronge le coeur de n'avoir osé prendre
Les jeux et les plaisirs de la jeunesse tendre.   [IV, p. 238]

Quant au passé, il fuit sans esperance
De retourner pour faire un lendemain,
Et ne revient jamais en nostre main.   [V, p. 356]

Du Bellay:
                    si l'an qui faict le tour
Chasse nos iours sans espoir de retour.   [I, p. 137]

Gilles Durant:
Du soleil la lumière
Vers le soir se déteint
Puis à l'aube première
Elle reprend son teint;
    Mais notre jour,
Quand une fois il tombe,
Demeure sous la tombe
Sans espoir de retour.[46]

The fact that time did not return was, it is true, no new discovery—the first poet of the *Roman de la Rose* had already formulated his awareness of this truth, but for him this realization was not one which affected the existence and activity of humanity. The sixteenth-century poets, whose thought showed in general little originality, doubtless took over the idea from the classical authors. It was, however, one which appealed to them, and they formulated it with new conviction and allotted it a place in their view of life, to which it was well suited. The Renaissance readily borrowed a good deal of material from outside sources, but in the main only that which was in harmony with its tenor of life and its ideals. This borrowing, therefore, does not throw doubt on the genuineness of its feeling for the unrepeatable quality of experience.

Events and periods that were unique and which would never recur were now observed and described as such. The case of the day exemplified this. The day was now no longer something general and indefinite, but it was a section of time which represented a singular and unrepeatable experience. Agrippa d'Aubigné makes us experience the day following the massacre of Saint Bartholomew as one we cannot liken to any other, as an *individuum ineffabile*. The day itself was actively and indissolubly bound up with its happenings. Its aspect was fatefully impregnated by the historic event. This sense of the uniqueness of the occurrences and experiences of the day enhanced the historical evocation.

> Voicy venir le jour, jour que les destinées
> Voioient, à bas sourcils, glisser de deux années,
> Le jour marqué de noir, le terme des appasts,
> Qui voulut estre nuict, et tourner sur ses pas:
> Jour qui avec horreur parmy les jours se conte,
> Qui se marque de route et rougit de sa honte.
> L'aube se veut lever, aube qui eut jadis
> Son teinct brunet orné des fleurs du Paradis;
> Quand, par son treillis d'or, la rose cramoisie
> Esclattoit, on disoit: "Voicy ou vent, ou pluye."
> Cette aube que la mort vient armer et coëffer
> D'estincellans brasiers ou de tisons d'enfer,
> Pour ne desmentir point son funeste visage,
> Fit ses vents de souspirs, et de sang son orage.[47]

The uniqueness and individuality of a period of time thus had two aspects. It was deplored as something negative in its non-recurrence, and yet at the same time the fact that it supplied a particular experience made it seem desirable. The awareness of being surrounded by impermanent things strengthened one's will to counteract impermanence and submergence in time, to wrest its treasures from it and in some sort to mitigate its irreversibility. The humanist hoped that posthumous fame would preserve him from oblivion. Time was an ever-present force on which as an ordinary individual one always felt dependent, and this feeling of dependence was strengthened when it was seen, through humanist eyes, as something to be overcome. A marked spiritual tension arose between one's sense of transitoriness and one's desire for self-perpetuation here on earth. In the imagination of the sixteenth century, time became an enemy of man, because it set bounds to his existence. It was a personified force which sowed seeds of dissention in the human mind. The idea of time as an autonomous personal deity was already gaining ground in Italy. Petrarch was the first to become painfully aware of the gulf between the sense of personal impermanence and the desire for survival in posterity. In his *Trionfi* the triumph of time, seen as an adversary and assassin of fame, was added to the triumph of love, chastity, death, renown and the eternity of God. Time as a philosophical concept thereby acquired its autonomous value, and became associated with another value, posthumous fame. Time appeared to the poets in the guise of a sun which rose and set with uncanny rapidity. When one's gaze was fixed on its sempiternal course, human divisions of time seemed as infinitesimal instants. Succession became almost a matter of simultaneity.

> I' vidi il ghiaccio, e lì stesso la rosa:
> Quasi in un punto il gran freddo e'l gran caldo.

Seen in this perspective, posthumous fame was only a matter of sooner or later. It was nothing absolute. For Petrarch, this consideration was no literary plaything but a realization at once joyous and painful. The power of time and its enigmatic countenance troubled him even in youth. He set his will to overcome time against his awareness of the irrationality of the future. The more uncertain the latter became, the more he clung to his hope of self-perpetuation by posthumous renown,[48] driven by a heroic sense of life which dared to take a stand against time.

The antithesis between renown and time lived also in the imagination of the Renaissance poets in France. With them, however, it manifested itself less in the form of pessimistic contemplation than as a lyrical motif in affirmation of posthumous fame.

The French did not place or visualize themselves outside the contest between eternity and impermanence, but impatiently, and perhaps blindly, made pretensions to immortality on their own account. This striving to endure through all time here on earth is discernible in France as early as the fifteenth century, when it took a particular form. In 1407, when the Duke of Orleans was murdered in Paris by the men of John the Fearless, it was decided that this deed should be kept fresh in human memory by the pulling down of buildings, the erection of monuments and crosses and the celebration, on a permanent footing, of mass in a special chapel to be built for this purpose.[49] Eternity, a concept hitherto confined to the Beyond, was invoked on the strength of a worldly, human affair. Earthly immortality, in the form of remembrance or fame, now became a goal to be attained.

The fact that noteworthy deeds were entered in the annals, and the desire to ensure that the heroes of contemporary history should occupy a lasting place in human memory, imparted a new significance to literature, albeit for the time being only to historical writings.[50] The medieval antithesis between the eternal and the temporal thereby lost much of its tension. The human, the vain, and the fleeting moment might be enshrined in eternity; the ephemeral and the nugatory might be made to endure. The 'amor di cosa che non duri eternalmente'[51] contemned by Dante was affirmed by the transference of the concept of eternity to temporal matters. Perpetuation of oneself and one's works was no longer something remote, lapped in the mysteries of what happened beyond the grave, but now lay within the reach of human will.

This mode of thought lent a dynamic quality to the idea of lastingness; it was not something that already existed, but had to be striven for and made real. By the fifteenth century the words *toujours* and *jamais* had become inadequate in themselves for expressing the new desire for lastingness. In Old French these words had implied rhythm and direction, but in the course of their development they increasingly became indications of extension. The old *jamais* had completely lost its future-orientated sense and now served to denote all temporal possibilities. In order to render the former connotation of the word, *à jamais* was now used; the prefixed preposition lent wings to the idea of duration, causing it to 'take off' and turning it from a state into a wish.

Roi René:
La somme de cent livres tournois de rente annuelle et perpetuelle pour dire et celebrer à jamais perpetuellement une messe basse.
[I, p. 87]

Saintré:
Sont venus en si hault honneur, que à tousjours en sera nouvelle.
[Ch. 3]

Paré:
Tous lesquels ont eu en si grand honneur, reverence, et recommandation les corps des trespassés, pour l'esperance de resurrection, qu'ils ont fort recherché les moyens, non seulement de les ensevelir, mais aussi de les conserver à jamais.
[III, p. 475]

A fin que l'espece demeure à iamais incorruptible et eternelle.
[II, p. 635]

Belleau:
Vous, ministres de ma victoire,
    En memoire
A iamais ie vous vanteray.
[I, p. 108]

De mille morts ie meurs quand d'une extreme envie
Ie desire à iamais luy estre serviteur.
[I, p. 190]

Un reproche eternel à iamais lamentable.
[III, p. 30]

Des Périers:
Sa renommée
    Nommée
En sera à tout jamais.
[I, p. 76]

It was Ronsard's thought that the secularization of the idea of eternity reached its highest development. He extolled poetry as something timeless. The poet's most important task seemed to him to be the perpetuation in posterity of individuals and values. There was only one possible means of escaping from the clutches of transitoriness: to be immortalized by the poet. It was he who held the monopoly of independence from time.

Je suis le trafiqueur des Muses,
Et de leurs biens, maistres du temps.
[II, p. 114]

Ne laisse pourtant de mettre
Tes vers au jour, car le metre
Qu'en toy bruire tu entens
T'ose pour jamais promettre
Te faire vainqueur du temps.   [II, p. 443]

In much of his verse Ronsard represented time as the adversary of man. Resentful of the power of transitoriness, one defined its boundaries at every opportunity and repulsed its encroachments. This reflected the sharpened sense of the sixteenth century for the varying degrees of permanence of things and values. In its eyes, time became a means of passing judgment. With this was associated the practice of gauging the worth of a person by the age of his family tree. The individual boasted of the tenacity with which his kindred had resisted the centuries. The recollection of one's remote ancestors gave one a sense of victory over time. It was not only veneration for antiquity that induced Ronsard to trace the genealogical tree of the French kings back to the legendary Francus, but also the new esteem for lastingness and for age *per se*. Thus Du Bellay considered the *Roman de la Rose* to be worth reading, not for its literary or other importance in itself, but because it was a memorial of the French language which 'merited honour on account of its ancientness'.[52] The powers of endurance or the impermanence of things were now constantly inquired into. The same Ronsard who had an affection for the passing moment and who advocated the philosophy of *carpe diem* was fond of using such words as 'eternal', 'eternize', 'permanent', 'immortal' and 'the hereafter'. Only that which resisted time was valuable and genuine. To point to the permanence of a thing was a new, humanistic way of praising it.

Ronsard:
Tousjours la vertu demeure
Constante contre le temps.   [II, p. 66]

J'ay juré de faire croistre
Ta gloire contre les ans.   [II, p. 68]

Et croy que c'est une richesse
Qui par le temps ne s'use pas,
Mais contre le temps elle dure.   [II, p. 114]

   Car cecy peut durer
Ferme contre le temps.   [VI, p. 204]

Le desir, au desir d'un noeud ferme lié
Par le temps ne s'oublie et n'est point oublié.   [I, p. 312]

Un dueil, que le temps n'a pouvoir
D'arracher de ta souvenance.   [II, p. 139]

Mais le sçavoir de la Muse
Plus que la richesse est fort;
Car jamais rouille ne s'use,
Et maugré les ans refuse
De donner place à la mort.   [II, p. 205]

Ainsi le cours des ans ta beauté ne fanisse
Ains maistresse du temps contre l'age fleurisse.   [IV, p. 224]

   Jamais le temps vainqueur
N'ostera ce beau nom du marbre de mon cœur.   [IV, p. 300]

Leur nom qui le temps surmonte.   [*Odes*, I, Ode 5]

Du Bellay:
Mon amour, qui vous sera sans cesse
Contre le temps et la mort asseurée.   [I, p. 100]

Mais ces escripts, qui son loz le plus beau
Malgré le temps arrachent du tumbeau.   [II, p. 266]

Such observations expressed an awareness that an indication of length of life was the most important thing that could be said of a person, a value, a sentiment or a thing. In phrases such as 'la maison eternelle'[53] or 'l'eternelle promesse'[54] duration was no random attribute, no property loosely associated with a thing but a quality which determined its nature and on which its innate value depended. This notion manifested itself in the language as a tendency to use an adjective in place of a temporal adverb. By this means the innate inseparability of a thing from time was emphasized.

Ronsard:
Au soir, à la tarde chandelle.   [II, p. 307]

Où le coq si tardif nous annonce le jour.   [I, p. 339]

Du Bellay:
Feuillette de main nocturne et journelle les exemplaires grecz et latins.   [*Deff. et Ill.*, ch. 4]

Time was now represented and felt as an integral part of things and essences. In the fourteenth and fifteenth centuries it was a completely external condition for all happenings, being only incidentally linked with events. In Middle French, temporal adverbs were joined with a verb of happening only in a random and non-binding manner. Time was not a property of things, but a circumstance. Now, however, the temporal position of an event became one of its essential characteristics. What was said of things in respect of time referred to their immanent qualities. When Ronsard used *tard* adjectivally he exemplified the spiritualization and introversion of the idea of time. The presence of time was now felt in everything. It was encountered at every turn, and invariably one of the two basic attitudes towards it characteristic of the century was adopted: one sought either to live through it as an experience or to defeat its effects by gaining fame in posterity. Both of these attitudes arose from a spiritual need—a desire to escape from transitoriness in some way. During the course of life this desire was satisfied by making time impart to an experience a degree of intensity proportional to the degree of quantity it denied. Outside life, in the hereafter, the name that lived on resisted time. We may discern in both these attitudes a profound anxiety with regard to transitoriness. It was less a fear of death, the dread which pervaded the Middle Ages, than a fear of annihilation. It was not a matter of wresting as much time as possible from death and destruction, but the issue was one of 'to be or not to be'. It was not a question of a greater or lesser amount of time in extension. The mentality of the Renaissance was concerned only with the alternatives of self-perpetuation or oblivion.

If, in this century, humanity averted its gaze from the future, it will be evident from what has been said that it was from the problems of the morrow that it turned away. The individual was disinclined to ponder on what was to come in a practical and sober manner. And yet his gaze was directed to the future nonetheless. He was preoccupied with what was remote, in time with later generations, posterity, the centuries to come, and posthumous fame. This was a markedly heroic attitude towards time. One was disinclined to deal with it in a petty manner and to obtain trifling advantages accruing from saving time, forestalling others and so on. A much more comprehensive claim was staked on the future.

This heroic attitude was adopted by the Renaissance even when its glances were retrospective. Unlike François Villon, who was unable to bridge the gulf between the past and himself, the Renaissance succeeded in doing so. In contrast to the Middle Ages, which were, so to speak, timeless in as much as they had no capacity for experiencing changes in time or temporal extension, the Renaissance possessed a ready and fluid ability to move backwards and forward in time. When one imagined oneself in an age long past, one had no feeling of overcoming distance. Time was not an historical reality, tragically separating the things and men of a former age from those of the present. It was not until the later Middle Ages that it gradually began to fade from consciousness as something cutting off the past; this meant that the setting up of intellectual links with bygone ages appeared as a conscious surmounting of an existing barrier. Communion with the figures of a past era seemed to be a reward vouchsafed to a heroic will that bridged time. In Italy, Petrarch had already endeavoured to overcome time *qua* the past, and in France Montaigne, Rabelais, Amyot and, perhaps most consciously of all, Ronsard did likewise.

> Mon grand Pindare vit encore,
> Et Simonide, et Stesichore,
> Sinon en vers, au moins par nom.   [*Odes*, I, Ode 16]

Men of intellect stretched out their hands to one another in both directions in time. This striving to assert oneself in time had its counterpart in the spatial sphere. The Renaissance constantly referred to the spatial as well as the temporal spreading of fame. 'Ever' and 'eternal' went hand in hand with 'everywhere'. The century which was introduced by the discovery of America was familiar with the ethos of spatial extension. It seems that no conceptual or linguistic means of conveying the worth of an individual or thing by indicating his or its effect and significance in space existed before the sixteenth century. Only on its lips did the terms *l'univers* and *le monde* acquire a force and vividness that enabled them to be used as yardsticks of the value of things. Robert Garnier caused his Augustus to besot himself with the conceit that he was, like Jupiter hurling a thunderbolt, 'broadcasting his words from one pole to the other'.[55]

And Ronsard sang:

> Ta fameuse renommée
> Qui doit voir tout l'univers.   [*Odes*, I, Ode 12, v. 59]

> Afin que son doux message
> S'espande par l'univers.   [Ode 14, v. 14]

In this epoch, expansion in space and penetration of time were equally balanced. 'La Renaissance et le Romantisme firent de l'expansion dans l'espace et de l'expansion dans le temps de vigoureuses complémentaires.'[56] But the Renaissance was still lacking in awareness of space and time as concepts of value. It did not yet realize that to make oneself felt in duration on the one hand and in spatial extension on the other were two different possibilities open to a person. Space and time were not yet discriminated or thought of as having unlike qualities. The antithetic natures of the two did not become apparent to humanity until later, when they impelled it to make a conscious decision: 'He who despises the acclaim of the crowd today does so because he seeks to live on in the minds of ever-changing minorities throughout generations—he desires to perpetuate himself in time rather than in space.'[57] But the centre of gravity of the desire for fame was placed by the Renaissance on time. When poets speak of themselves, pass judgment on their merit themselves, and talk of things whose value they champion with their entire consciousness and power of poetic expression, they are striving for temporal, not spatial, absoluteness. They feel that true being exists in duration.

The flux and course of time

Whereas, in the declining Middle Ages, duration and transitoriness were apprehended chiefly in their visible effects, and in the form of differences resulting from a comparison of the earlier with the later, and of former times with the present (two distinct images), there now existed a more acute sense of the passing of time itself, of the running, rushing, flowing, flying and gliding movement of time. Life and death, youth and old age were no longer set against one another *en bloc*, but youth and age were now experienced as extended in time. The points marking the beginning and end of a temporal phase were now overshadowed by the experience of living through the phase itself. Duration, so to speak, had become changed from a thing into an event, and from a concept into an object of sense-perception. The medieval idea of time was knot-like, and that of the Renaissance sinuous and melodic. To the Middle Ages time appeared as a knotted thread, of which only the knots were really visible. One was then conscious of time only at the beginning or completion of a process. In the fairy-tale, for example, time became real and perceptible only at the moment when the spell took effect or was broken—after Sleeping Beauty awoke from her hundred years' slumber, to cite an instance. The running of the thread of time between events went unnoticed.

In the consciousness of the Renaissance poets, however, time was a continuous thread; time was present in every moment, constantly flowing and singing. The 'knot image' disappeared from their minds; it was regarded by Rabelais and Montaigne as inflexible, inhuman, barbaric and medieval. It was no effect of chance that the image of the smoothly gliding thread was now applied to time:

Ronsard:
Par le fil d'une longue espace.   [*Odes*, I, Ode 10, v. 591]

Belleau:
Et couler doucement le fil de nos beaux iours.   [III, p. 179]

    Ceste ieune bergere
Qui faisait la mesnagere,
Noya le fil de ses iours.   [III, p. 146]

La vanité qui suit le fil de nostre vie.   [III, p. 197]

This image is based on the ancient idea of the three Fates who spun the thread of life.

The Renaissance represented temporal duration in language in various forms. It was seen either as the ceaseless flowing of a liquid—amorphous, unbroken, undivided, as pure transition; or as an articulated process, as something regularly repeated, as the growth and death, the coming and going of the same entities. The former view of the course of time represented duration as something beautiful; the latter portrayed it as something exalted.

The designation of time as something flowing was a reflection of sense impressions, whereas the idea of a 'striding' process, of the coming and going, the dawning and fading of periods of time, generally implied a more pronounced conceptual content. The flowing of time was an object of perception, in which the attention of an individual might be focused on the smallest span of time, on a single instant. The course of time, however, extended throughout the centuries. The Renaissance was familiar with both broad and narrow temporal perspectives. The poets were conscious of the most striking attribute of time, its flux. The units of time readily lost their fixed quantitative value when the words *glisser* and *couler* were applied to them; their limits became indefinite; they became acoustic rather than visual impressions, possessing direction rather than extension.

Ne sens-tu pas que le jour se passe?   [II, p. 150]

asked Ronsard. Humanity acquired thereby a new sense, enabling it to enter into a more profound and intimate relationship with things and allowing it to perceive their true nature as an inward and musical growth and decline.

Ronsard:
Le temps subtil à couler et passer.   [I, p. 371]

    La mémoire
Des siecles ja coulez.   [V, p. 18]

Au soir quand le jour est coullé.   [V, p. 262]

Mais tout ainsi que l'onde, à val des ruisseaux, fuit
Le pressant coulement de l'autre qui la suit;
Ainsi le temps se coule, et le present fait place
Au futur importun qui les talons luy trace.
Ce qui fut, se refait; tout coule comme une eau,
Et rien dessous le ciel ne se voit de nouveau.   [V, p. 248]

L'âge qui glisse.   [II, p. 133]

L'âge glissant qui ne s'arreste.   [II, p. 365]

Du Bellay:
Bref ne laisser couler, soit de iour soit de nuict,
Une heure sans plaisir.   [II, p. 484]

Etant ja coulé de mon aage le temps le plus apte à l'etude.
                                                        [I, p. 71]

Belleau:
C'est toy qui retiens en bride
Des heures le glissant pas.   [I, p. 81]

The Latin poets of the classical age had used the same forms
of expression to convey the passing of time. The image they were
most fond of using in order to lend a pictorial quality to time was
that of the quiet flowing and washing of water:

Ovid:
Labitur occulte fallitque volatilis aetas.   [*Met.*, 10, v. 519]

Cuncta fluunt, omnisque vagans formatur imago,
Ipsa quoque assiduo labuntur tempora motu,
Non secus ac flumen: neque enim consistere flumen,
Nec levis hora potest, sed ut unda impellitur unda,
Urgeturque eadem veniens, urgetque priorem;
Tempora sic fugiunt pariter, pariterque sequuntur,
Et nova sunt semper, nam quod fit ante relictum est,
Fitque quot haud fuerat, momentaque cuncta novantur.
                                            [*Met.*, 15, vv. 178–85]

     Eunt anni more fluentis aquae;
Nec quae praeteriit, iterum revocabitur unda,
Nec quae praeteriit, hora redire potest.
                                    [*Ars amatoria*, 3, vv. 62–4]

The fact that the image was similar with both the Latin poets
and the French doubtless resulted from the literary dependence
of the latter on the former. If we speak of the flowing of time,
it is probable that classical influences were in part responsible for
the creation of this metaphor. It may be that the equation 'time
= water' was not one to come to mind without some influence
or other. It was not an association that generally suggested
itself—that is, it did not spring to all minds at all times with
equal readiness. The course of time had already been compared
to the flowing of water by Guillaume de Lorris,[58] and it is likely
that he borrowed the image from antiquity, to whose thought it
was closer, because for it the running of water from the *clep-
sydrae* stood for the flux of time. The medieval symbol of time
was not the running of water but the firmly articulated and in-
cisive chimes of the striking clock, as described by Dante.[59] The
word *couler* conveys an auditory and visual impression of time as
something round, soft and uniform. The following are the first

examples of the application of the word *couler* to time, taken from a fifteenth-century author:

Chastellain:
Le temps couloit tousjours.   [III, p. 81]

Comme le temps va tousjours coulant.   [V, p. 254]

In the century of the Renaissance time was generally apprehended as a flux and movement. As a result of thinking in units of time and in consequence of an ever clearer orientation in the given temporal order, the idea of time in the Middle Ages had become so inflexible that it was pictured as a fixed structure, within whose confines one moved. Time was the measure of human action; the moving and transient world was transformed to a calm, regular plain, the plain of time.

This situation was completely reversed by the poets of the Renaissance. Humanity now occupied a stable emplacement and time flowed past it. Such an egocentric view constituted a more natural relationship to time, and one more in keeping with human perceptions, for the immediate and uninfluenced consciousness always sees itself as something persisting. The expanses of time now no longer marched through, but instead flowed rapidly from the unreality of the future through the reality of the moment and into the new unreality of the past. To this way of thinking, units of duration no longer had a fixed extension but were waves whose length was a matter of indifference. This hesitation between the egocentric and the static views of time is something generally inherent in human thought. We shall express no opinion as to which position is the more plausible, but it seems to us that the one is the expression of a subjective, and the other of an objective, mental attitude. Even the assertion that time 'passes' or 'flows' arises from the need to render a process, inaccessible to the senses alone, imaginable. The idea of time as something flowing is only a makeshift of the imagination, and which yet is more pictorial and more consistent with one's inner sense of time than is the view of time as a rigid structure.

The passing of the days, weeks and years also appeared as an eternal cycle, an ever-repeated coming and going. This has become a very general—indeed, utterly banal—idea with us;

but in sixteenth-century poetry it was something new. The course of time was experienced, as it never had been before, as an unvaryingly regular rhythm: coming and going, to and fro, day after day. The ear was receptive to the music of the happenings of the world, and of the course of the years and centuries.

Ronsard:
Toy, comme second pere, en abondance enfantes
Les siecles; et des ans les suites renaissantes,
Les mois et les saisons, les heures et les jours
Ainsi que jouvenceaux jeunissent de ton cours,
Frayant sans nul repos une orniere eternelle,
Qui tousjours se retrace et se refraye en elle.   [V, p. 143]

Within the context of eternity, human life shrank to a single day. The course of time was something so exalted that the mind which was attuned to this exaltedness adapted itself to the course of time and subordinated itself to the rhythm of nature. Since the individual had now come to understand the microcosm and macrocosm of creation, he perceived also the small and the great in their temporal context—the activities of the day, the passing of the quiet hours on the one hand, and the vast framework of time on the other, within which his own life seemed to dwindle to nothing. It was true that he felt emancipated from the monastic subservience to time, which no longer had any significance for him, but he was nonetheless subject to certain bonds from which he could not escape. He was an element of nature, a prisoner of the alternating cycles of day and night, summer and winter, work and rest, seed-time and harvest, joy and grief. There was a set time for everything:

Belleau:
Temps de naistre en ce monde, et de mourir aussi:
Temps de prendre plaisir, et de prendre souci:
Il y a temps prefix et certaine ordonnance
D'ensemencer la terre, et cueillir sa semence,
De planter, d'arracher, de tuer, de guarir,
De ruiner le vieil, et de nouveau bastir;
Temps de pleurs, temps de ris, de ioye et de tristesse,
De sauter, de gaudir, de se mettre en liesse:
Temps de ietter la pierre, et temps de l'amasser:
Temps propre d'embrasser, et temps de s'en passer:
Temps d'acquerir les biens, et temps de les despendre:
Temps de cueillir les fruits, et temps de les espandre:

Temps de coudre et descoudre, et temps de dechirer:
Temps propre de se taire, et temps propre à parler:
Temps de haine et d'amour, temps de paix, temps de guerre.
Qu'a l'homme davantage en ceste basse Terre,
Suant et travaillant, entre tant d'accidens
Qu'il prend sous le Soleil, que le cours de ces temps?

[III, p. 174, *Discours de la Vanité*]

The commonplace that there is a suitable time for everything, and that this is a necessity imposed by the course of nature, is here set forth in the light of an essentially new realization. In the above excerpt it is not the content of the words that matters but the emphasis placed on them. After the later Middle Ages had sloughed off their supine passivity in regard to the natural temporal order by adopting a flexible attitude towards circumstances prevailing at the time, a poet of the Renaissance could likewise see human destiny and preoccupations as a rhythmic process and a periodic cycle, bound to the elemental conditions of human existence. Everything one did, everything that befell one and everything that happened outside one's own existence were the effect of time in its passing and formed part of a vast synthesis. This fact was ever present, to a greater or lesser extent, in human consciousness. The individual was constantly reminded of the fact that the stream of time was bearing him away with itself, and that aeons of time had preceded, and would follow, him. Time did not stand still. It flowed on as a *cours*,[60] *course*,[61] *carrière*[62] or *pas*.[63]

All happenings of a particular nature had their particular temporal extension, prescribed by nature. Every process was a natural unity, the end of which was determined by its beginning. It resembled a step taken by time. The individual must not interrupt the course of natural events, but must understand the rhythm and duration peculiar to changes and states. Montaigne's dictum on the duration of illnesses reflected the general tendency of his century to fall in with the eternal laws of time and with its inexorable course:

Elles ont leur fortune limitée dès leur naissance, et leurs jours. Qui essaye de les abbreger imperieusement, par force, au travers de leur course, il les allonge et multiplie, et les harselle au lieu de les appaiser.   [III, 13, *Var.* 1595]

Time marched on at its own pace; one recognized this pace and was willing to adapt oneself to it. The idea of life and growth most faithfully reflected the belief that everything had its natural life span. The sharp distinction drawn by the Middle Ages between human beings and plants became blurred. Both lived, grew, blossomed and withered. Man was comparable to a plant, for he was subject to the same development and changes as it was. Comparisons and parallels drawn from the vegetable world are commonplace today. 'We apply all images associated with the growth and death of plant life to the human condition; just as leaves turn yellow, flowers wilt, and trees wither, so it is with our body.'[64] These images, however, had their genesis. Several new turns of phrase originating in the sixteenth century made it plain that human life was regarded for the first time as a process of growth resembling that of plants. Terms related to plant life were applied to man, and contrariwise. In Ronsard's famous *Ode à Cassandre* the words 'bloom', 'green' and 'wither' were used in this new application. In one of his quatrains Guy de Pibrac described man as a 'plante divine'.[65]

Such comparisons and metaphors suited the current predilection for equating the year and the seasons with the ages of life, and for speaking poetically of the 'spring of life'; this mode of thought, like the practice of drawing parallels with plant life, did not come about overnight. Not only did the course of human life become identified with vegetable growth, but historical developments also were viewed in the same light; in the mind of one whose glance was turned to the past they appeared as completely subject to natural laws. Before those who had misgivings as to the future development of the French language, and who doubted whether it was susceptible of perfection, Joachim du Bellay placed the image of the tree, which required due time in which to grow, flower, bear fruit and die.[66] The same poet also spoke of the 'growth' of philosophies.[67]

A year in the life of an individual is equivalent to a century in that of peoples and cultures. The poets of the Renaissance could not dispense with the word *siècle*. It formed part of their basic vocabulary. It became a commonplace to think in terms of centuries. The word, moreover, was not associated with any particular clarity of outlook. Its use was evidence more of a need for exalted sensations and of a lofty experience of time rather than for temporal orientation. The passing of the centuries and the vastness of eternity intoxicated the individual, inducing in him a feeling of giddiness with regard to time, and causing his life and his century to seem as nothing. Viewed in the light of this new sense of time, history ceased to be a collection of memorabilia, practical lessons and tableaux, but instead became a matter of grandiose destinies. Thus history appeared to Marguerite de Navarre as a devastating succession of ascending and declining world empires;[68] Du Bellay saw Rome, with its former grandeur and present decay, in the same light. The stones spoke to him of the sublimity of time.[69] Ronsard wrote an ode in the same vein.[70] The panorama of world history inspired a feeling of modesty. The course of time had something majestic about it. Although one was aware of the fact that time, as an inner experience, was irreversible and irrecoverable, it was readily seen as a grandiose external process, recurring in an eternal periodicity. Days, years and centuries turned about in a circle, ran their course and returned. The music made by the year as it followed its lengthy cycle was heard. This view of the course of time as an astronomical machine was, however, a living one. Everything passed away, but everything returned in a new form.

Belleau:
De là se retrame le cours,
Et l'ordre qui roule tousiours,
Des corps que ceste mesnagere
Nature défait et refait,
Tant seulement change le trait
Et l'air de l'image premiere.   [III, p. 141, *La Pierre Lunaire*]

In the same poem the word *renaissance* occurs, applied to this eternal flux of the natural and the human order of things:

De la mort vient la renaissance.   [III, p. 141]

'Recurrence', 'return' and 'rebirth' were words which caused human existence and the course of human life to appear as an element of a vaster cycle. The human, the astronomical, the vegetable and the historical formed part of a great rhythmic unity, the presence of which was felt rather than indicated. By virtue of its unwavering persistence, this rhythm imparted stability, continuity and beauty to the world, and in the human mind it inspired at once awe and confidence. In the succeeding century, corresponding to the classical era in France, the sense of the sublimity inherent in this recurrence was strengthened, and manifested itself as a life-moulding force through the majestic temporal order which the Sun King imparted to his own existence and to his circle.

The intellectual watershed

In the century of the Renaissance, French intellectual history was marked for the first time by the realization that an old order was fading and that another was coming into existence, an order which called for a changed attitude to life. It was realized for the first time that a choice had to be made between rules, compulsion, subservience and compliance to external circumstances and relations on the one hand, and freedom, joy of living and recognition of the individual on the other. During this epoch humanity immediately became aware of the important part played by time in this choice. That dictum which the poets of the Pléiade transmuted into personal experience, and which they advocated by the general drift of their poetry rather than by conceptual explicitness, was unequivocally uttered by Rabelais: the individual may be recognized by his attitude to time. One may differentiate between persons who live, and desire to live, by the clock, and those temperamentally disinclined to subordinate their lives to it. The clock was a symbol of a view of life which was now discredited. A rift opened between the person who drew up, and adhered to, a timetable, and the sovereign freedom of the devotee of the new way of life. Later, in the Romantic era, when people turned away from the old ideal of life and adopted a new, the clock became the symbol of philistinism. Tieck satirized it in his prologue to *Oktavian*. But that wise citizen Goethe, in his *Wanderjahre*, let clocks be set up everywhere in the 'pedagogic province' as an admonition to make good use of time.

The attack led by Rabelais on the clock-bound life was no mere incidental affair or chance manifestation of his likes and dislikes, no mere fancy which might be dismissed or imagined as taking a contrary form but was the necessary expression of a basic position which he inevitably took up. His answer to the question of what attitude one should adopt towards time is essentially more important, and tells one infinitely more about him than information concerning his favourite dish or his literary tastes. Time was no institution with regard to which one held varying opinions, but a general condition of human existence. In the intellectual sphere this condition, to which all were subject, might be passed over in silence or overlooked, but in practical life one must come to terms with it, that is to say, one must do something with it and take a decision about it. The Renaissance explicitly recognized this necessity, because it saw time as a basic phenomenon which could not be escaped.

It is idle to ask oneself whether Rabelais' confession of faith on the subject of time is to be taken at its face value or with a grain of salt. There is no doubting the significance of his challenge, and to this we must now turn. The thought of the day being divided by the chimes of the clock conjured up all manner of unpleasant associations in the mind of the author of *Gargantua*. His thoughts were not confined to the alternatives of subservience to the clock and freedom and unrestricted experience in time on the other, but lurking behind these alternatives he perceived the human types who adopted the one or the other. The inferior, the incompetent, the withered, the hidebound, the dependent, the narrow-minded, the wretched, the monks and the scribes, all who sat in monasteries and stuffy offices, the pen-pushers and those obliged to keep regular hours with an emotionless face, all those who were slaves of time instead of realizing it as an enjoyable experience, incurred his disapproval. Against this workaday servitude he set his holiday view of things, in which the moment was an end in itself. Those things done by one bound to the clock were done for the future. To this turn of mind, the present had to be turned to account for the sake of the future. Rabelais restored the present to its rightful place. Here we see not only two points of view, but two worlds, set against one another. The awareness of the new view of time often expressed itself as a mocking of the old attitude towards it. The mind of the Renaissance found this straitjacketing of experience by a lifeless temporal order not only unnatural and unfree but also ridiculous. This supplied the humorists with a new field of activity and new targets.

The relationship between time and humour had been discovered in the fifteenth century. In the *Cent Nouvelles Nouvelles* fun was poked at a country priest who had almost forgotten a feast day. This was the mocking attitude of the order-loving citizen who knew his calendar. In *Petit Jean de Saintré* an abbot emancipated himself from the clock in the most arbitrary manner: in order that he might have his mid-day meal at once, he put the clock forward an hour and a half so that it showed twelve o'clock.[71] This incident showed a new comic turn. Rabelais parodied the ponderous formality of dating because it lacked the living content that filled time. It was a shell without a kernel. He made the process appear as a meaningless gesture, the making of senseless marks on paper. In his works dating was overdone in comic seriousness. External time was not to be taken seriously.

> En icelle [année], les kalendes furent trouvées par les breviaires des Grecs. Le mois de mars faillit en quaresme, et fut la myoust en may. Au mois d'octobre, ce me semble, ou bien de septembre (afin que je n'en erre, car de cela me veulx je curieusement guarder) fut la sepmaine tant renommée par les annales, qu'on nomme la sepmaine des trois jeudis: car il y en eut trois, à cause des irréguliers bissextes, que le soleil bruncha quelque peu comme *debitoribus* à gauche, et la lune varia de son cours de cinc toises, et fut manifestement veu le mouvement de trepidation on firmament dict Aplanes.   [*Garg. et Pant.*, II, ch. 1]

Rabelais wrote in the same vein of sense or nonsense in his account of Pantagruel's birth:

> Gargantua, en son eage de quatre cens quatre vingtz quarente et quatre ans, engendra son fils Pantagruel.   [II, ch. 2]

> En icelle année fut seicheresse tant grande en tout le pays de l'Africque que passerent XXXVI mois troys sepmaines quatre jours treize heures et quelque peu d'advantaige sans pluye, avec chaleur de soleil si vehement que toute la terre en estoit aride.   [*Ibid.*]

The petty-minded scribes dated, chronologized, calculated and divided time, and furrowed their brows as they predicted leap years, worked out biographical dates and established ecclesiastical feast days. Rabelais laughed his giant's laugh at it all. Life must not be measured, but weighed according to its content and value. The author of *Gargantua* was the doyen of all later wits and jesters who mocked at objective time. His humour was somewhat tendentious here. He ranged himself quite consciously on the side of true time, where all calculations and fixing of dates had no meaning.

Greater and purer geniuses than he, who lacked his aggressive attitude, made use of these contradictory views of time in order to throw light on the many different constructions placed on the world and life and to bring out the differences between individuals of various types. In La Fontaine's fable 'The shoemaker and the man of means' the varying interpretations placed on time by the two figures elucidated their respective characters. The poet made the two different temporal orders in which the shoemaker and the *rentier* lived contrast with each other in a most impressive way. In doing this he was concerned less with demonstrating technically how their views of time differed than with indicating how this divergence was bound up with varying emotional attitudes.

The most outstanding example of a character sketch based on two contrasting views of time was that drawn by Cervantes in *Don Quixote*. Don Quixote had a romantic and chivalric, and his servant Sancho Panza a banal, conception of time. The overexcitable hidalgo, who had lost touch with the factual conditions of life and with others, and who saw everything through the distorting lens of knight-errantry, also distorted the normal temporal order by robbing the measures of time of their factual and unequivocal meaning and investing them instead with an abstract, symbolic and nebulous significance. He was out of his depth in the everyday temporal order in which his manservant was at home. Hence they failed to understand each other, and the wavelengths of their respective lives were much less well attuned than those of others. When his master wished to hear another song from the musical goatherds Sancho desired to sleep, because night had fallen.[72] Don Quixote spoke of the six days in which Sancho might be king, and the latter took him at his word. In another episode, when the two adventurers had recovered consciousness after being half beaten to death by the Yangueses, Sancho, after expressing his desire for the magic potion of Fierabras, received the following reply:

Mas yo te juro, Sancho Panza, á fe de caballero andante, que antes que pasen dos días, si la fortuna no ordena otra cosa, la tengo de tener en mi poder, ó mal me han de andar las manos.

To which Sancho answered:

¿En cuántos le parece á vuestra merced que podremos mover los pies?[73]

In the mind of the knight, time was devoid of its natural course and divisions; he did not see it as an organic whole. Temporal possibilities were therefore unlimited. Anything liable to happen at all—and this was a great deal—might come to pass at any time. Don Quixote had forfeited the most elementary capacity for finding his bearings in the everyday scheme of time.

It was natural that the new evaluation of time brought about by the Renaissance should cause the human life span to appear in a different light than formerly. The later Middle Ages, haunted by the spectre of death, desired long life above all. Time was evaluated by the yardstick of its objective length rather than by that of the experience it provided. The bourgeois world attached no differentiated value to the various ages of life, so far as intensity of experience was concerned. Decrepitude had the same rights as exuberant youth. According to this view, one's forces should not be expended in one particular age of life, but should be equally apportioned over one's entire life span. This was no theoretical and philosophical doctrine but a precept expressed by the author of the *Roman de la Rose* when he spoke of those evil ways which shortened life.[74] The French author of the *Curial* said of a life of moderation that

de telle vie s'esjouist nature, en telles petites cases vit elle longuement, et petit a petit s'en va jusques a plaisant vieillesse et honneste fin.[75]

In these testimonies we may see more than the expression of the instinct for self-preservation, implanted by nature in everyone, This instinct was raised, in a quantitative–temporal form, to a general ideal of life characteristic of the declining Middle Ages. To lead a long life was not only a matter of practical good sense but also a morally desirable end. Christine de Pisan, as she imparted valuable precepts to her son, placed this goal before his eyes:

Veulz tu vaincre et long temps durer?[76]

It was the ideal of the Old Testament: 'Mayest thou fare well and live long in the country which the Lord thy God gave thee.' But it was also the goal of the medieval mind, for which there was an ideal life span: eighty-one years. Plato lived to this age, as Christ would have done had He not died a Redeemer's death. To attain this age was a distinction, a privilege and a reward. A long life here on earth was a consideration which began to carry more weight than it had in the high days of monasticism and chivalry. A new way of praising or condemning a thing or activity was to point to its possible effect on the length of one's life. References to this occurred in unexpected places. Martial d'Auvergne wrote in praise of the pleasant life and agreeable pursuits amid the springtime countryside:

> Car petiz et grans
> En vivent plus d'ans
> Selon leur desir.[77']

Present enjoyment was here linked with the idea of long life. Deschamps also thought of it. Those who turned night into day and inverted the natural order could never reach a ripe age. He exclaimed to them:

> A vostre mort courez plus que le cours:
> Trop me merveil comment vie vous dure.   [II, p. 132]

In the fifteenth century a slackening of the strictly moral view of life became noticeable. The intrinsic value of a life of moderation disappeared. A more concrete objective took the place of the clear conscience and the saintly existence: good health and long life. In a morality play of the later Middle Ages, *La Condamnacion de Bancquet*, its author attacked gluttony exclusively from the point of view of health. If the watchword was formerly 'A life of moderation is a godly life', it was now:

> Je croy que l'homme qui vouldroit
> Faire ung repas tant seullement,
> Tousjours santé garder pourroit,
> Et si vivroit plus longuement.[78]

Long life was a desirable end in itself, the value of which was now rendered independent of all other values. Contrasts between life and honour, life and happiness, life and love, were unknown before the era of the Renaissance. Life was something to whose value humanity had just opened its eyes, something to which it devoted itself wholeheartedly in greater or lesser disregard of other values, as is always the case when the intellectual horizon becomes widened or changed. This feeling for life is, of course, truly discernible where spiritual reality was not concealed or distorted by 'literature'. The literary *taedium vitae* of a *rhétoriqueur* such as Charles d'Orleans, who wished to die because his beloved was no more, was too fluid in form to carry much weight.[79] In connection with another matter the same poet complained in his allegorical language of old age, the concomitants of which were besetting him, but he did not make any outright condemnation of *la vieillesse*. It formed an integral part of the medieval view of life; 'c'est le cours de nature'.[80]

It is noteworthy how markedly the Renaissance differed in this respect from the centuries which preceded it. The watchword for life was now 'Short and good'. The idea of life acquired a new meaning. The widsom of Seneca's dictum that 'to live' and 'to exist' were not the same thing, was appreciated.[81] Only when existence was fully lived as experience did it count as life. Its extension in time became a secondary consideration. The 'existence toute en longueur', as Marcel Proust called it, was discredited. Life was weighed, not measured. The *carpe diem* was the most eloquent expression of this feeling for time, which sought to apprehend duration as intensity of experience. Even if the length of time one had yet to live was independent of one's will, one should choose to live consciously and in depth rather than in superficial extension. In this context 'to live doubly' did not mean 'to live twice as long' but 'to live with twofold intensity'. No one saw the difference between these two views of time more clearly, or made a more decisive choice between these alternatives, than Montaigne. He said of life that

Il y a du mesnage à la jouyr: je la jouys doublement des autres, car la mesure en la jouyssance depend du plus ou moins d'application que nous y prestons. Principallement à cette heure que j'aperçoy la mienne si briefve en temps, je la veux grossir et estendre en pois; je veux arrester la promptitude de sa fuite par la promptitude de ma sesie, et, par la vigueur de l'usage, compenser la hastiveté de son escoulement; à mesure que la possession du vivre est plus courte, il me la faut rendre plus profonde et plus pleine.   [III, 13]

Of all the figures of French intellectual history it was Montaigne who stated his views on the subject of true duration in the most original, the most artistic and the most sincere manner. He was unable to find sufficient scope adequately to express his attitude towards, and treatment of, time. For all we know, he was the first Frenchman, and perhaps also the first person in modern times, to draw such a sharp distinction between the two alternative attitudes towards time—that based on depth and that based on superficial extension. In every instance and at every moment Montaigne was aware that he had to choose between various modes of looking at and dealing with circumstances, and that his every decision made him the exponent of a particular view of life and of time. His attitude was a conscious one, not merely on occasion, but consistently so. This attitude is apparent from his treatment even of seemingly trivial subjects, and moreover it was acknowledged by him. His detached manner of writing must not be allowed to blind one to the fact that not only was he drawing a portrait of himself, Michel de Montaigne, with all his idiosyncrasies, but also that in doing so he was decisively expressing his view of life. He dealt many blows—less resounding, perhaps, than those administered by Rabelais, but they struck home none the less. In his reflections on the subject of leisure activities he marked himself sharply off from his fellows. The attitude of a person towards time incited Montaigne to judge his mind, and to weigh up and appraise his mind. This manner of appraisal caused the phrase 'life of poverty'—used by the poet Belleau in respect of a miser who did not dare to live—to appear as an epithet of spiritual evaluation.[82]

A person's time—that is, his life—was most truly poverty-stricken if it was devoid of inherent value as an experience, if it lacked the qualities of depth and intensity. The ethos of the decisive renunciation of the wretched, vacuous life, having only extension, was championed in truly poetic terms by Louise Labé:

Ie ne souhaitte encore point mourir.
Mais quand mes yeus ie sentiray tarir,
Ma voix cassee, et ma main impuissante,

Et mon esprit en ce mortel seiour
Ne pouuant plus montrer signe d'amante:
Priray la mort noircir mon plus cler iour.[83]

There existed earlier a form of yea-saying to death, but one which sprang from different sources. The latter Middle Ages were familiar with the spirit of *taedium vitae*—opposed by the Church—and of resignation to death. It was the awareness of the emptiness of life which called for its shortening. 'It was the *homo desolatus* who, in his disquiet and agitation, desired a state of repose free from fear and tribulation.'[84] This was the spiritual antithesis of the view held by the Renaissance, which saw in timely death a means of elevating life to a valuable and meaningful whole. Time, in and for itself, was nothing unless it was turned into something. It was required to serve a purpose, because the image of an ideally filled lifetime existed in the mind of humanity. He who spoke of 'time' said very little, for it had to be informed by human agency if it was to acquire significance.

The beauty of the day

In the light of what has been said of the conception of time of the Renaissance, it is not surprising that the day acquired a significance different from that which it had possessed in former centuries. Mention of a return to a natural order of time makes one think first of all of the natural day, an elemental entity whose influence on the course of life was so overriding that no one could evade it. In the later Middle Ages the day had been turned into an organizational and official unit; now it once again became something to be lived through as an experience, and whose beauties were sung by the poets. One was now concerned less with what the day brought and offered than with the manner in which it presented itself to the eyes, receptive to natural beauties, of the poets. A typical phrase introduced in the period of Middle French was *bon jour*, which implied the satisfactory completion of one's undertakings and business, a fortunate outcome to some affair, happiness and so on, in so far as one's attitude to time in those days was indeed an interested and active one. It is astonishing how little the Frenchmen of the thirteenth, fourteenth and fifteenth centuries had to say of the day itself, and of its luminous course. It is possible that its duration and brightness were separate objects of perception, and that in imagination they were no longer apprehended as going to make up the day.

Clément Marot was the first to rediscover the day as a source of inspiration for lyric poetry. Even so, his descriptions of the sunrise were embroidered by mythological references to 'Phœbus' and 'Aurora'. For him, the day was not a pre-eminent poetic motif. Marot did not share the true spirit of the Renaissance, which once more took a naive pleasure in the delights of the daytime. If, in the later Gothic era, it was the 'goodness' of the day which stood in the forefront of consciousness, it was now its beauty which did so. This change reflects, in our opinion, the transition from the outlook of the characters of the *Cent Nouvelles Nouvelles* to that of the humanity of the Renaissance, from interestedness to naturalness, from the bourgeois to the poetic turn of mind. The phrase *bon jour* was a piece of everyday banality. It indicated the future-orientated nature of the thought of the later Middle Ages. The 'good day' was a wish whose fulfilment lay in the future. The 'beautiful day', on the other hand, was a present experience which was apt to monopolize one's attention and which one readily enjoyed.

Belleau:
Dont revenant à moy, ie m'osté ceste envie
De iamais travailler, voulant tramer la vie
D'un homme de plaisir, savourant le beau jour.   [III, p. 173]

Ronsard:
Le jour qui plus beau se fait,
    Nous refait
Plus belle et verte la terre.   [I, p. 221]

Sous le beau jour qui s'allonge.   [*Odes*, I, Ode 19]

Au plus beau jour d'esté.   [I, p. 94]

Il est temps de laisser les vers et les amours
Et de prendre congé du plus beau de mes jours.   [I, p. 367]

Eschauffez d'un beau jour.   [I, p. 371]

    Cependant ce beau jour
Nous doit apprendre à demener l'amour.   [I, p. 375]

Quand le ciel eut allumé
Le beau jour par les campagnes.   [II, p. 228]

Aussi tost que l'Aube aux doigts de rose
Aura versé le beau jour de son sein.   [IV, p. 312]

By virtue of this new way of experiencing it, the day expanded, becoming so comprehensive that it was almost always referred to in the singular. The old *carpe diem* had likewise contributed to this expansion. One's attention was concentrated on one day only, and this single day was the sole subject of discourse, because the morning lay far behind. This day, with its present radiance, was a thing, a world, a life in itself. It is not surprising, therefore, that the equation 'day = life' was of frequent occurrence. The Renaissance poet, writing in the manner of Petrarch,[85] used the image of the day to symbolize the brevity of life. Life, like the day, was a time of light, a joy, a feast that must be celebrated before night fell. Yet it was also a phantom, as intangible and obsessive as its beauty. 'Our day' was 'our life'. As Ronsard did in his poem 'A Cassandre', other poets (Des Périers, Baïf, Du Bellay) likened it to a rose, whose beauty lasted but one day. In the verse of the poets it became a symbol of the day, and consequently of life and transitoriness. This was the ancient symbolism of the rose.[86] The Middle Ages did not apprehend the short life of this flower as a real, pictorial image. This brevity might, indeed, become the symbol of enduring joy.[87]

The day, however, was not merely a replica of life. It was something so beautiful that it might symbolize all possible values. As a result, the word *jour* became increasingly literary, steadily losing its everyday, practical meaning. In this way the idea of the day assumed a predominant place in language and consciousness. True, poets of all ages have expanded the narrow meaning of the word, but it was the particular achievement of the Renaissance to have made full use of the poetic possibilities inherent in the term. In this matter there can no longer be any question of influences and the borrowing of poetic modes of expression. The 'beautiful day' and 'the day' became synonymous with beauty in general. One who said *le beau jour* used the adjective as a gratuitous and decorative attribute. The day was always beautiful. Neither was it merely this or any other day, but one which provided an idea of experience, raising the empirical day to a prototype free from all contingencies of actual life. Only the image of the day could serve to express the admiration which filled one, or the joy and pleasure which one felt. *Le beau jour* became a synonym for 'a thing of beauty':

Ronsard:
Mais, tout ainsi que les flambeaux
Ou du Soleil ou d'une estoile
Tout soudain reluisent plus beaux
Après qu'ils ont brisé leur voile:
Ainsi, après ton long sejour,
Tu nous esclaires d'un beau jour.   [II, p. 37]

O terre fortunée,
Des Muses le sejour,
Qu'en tous les mois de l'année
Serene d'un beau jour!   [II, p. 155]

   Ains que mes yeux
Quittent le beau jour des cieux.   [II, p. 348]

L'art, la nature et les astres encore,
Les elements, les grâces et les dieux
Ont prodigué le parfait de leur mieux
Dans son beau jour qui le nôtre décore.   [I, p. 4]

Of the eyes of a beloved woman, Ronsard said:

De leur beau jour le souvenir m'enflame.   [I, p. 53]

Du Bellay:
Et d'un beau jour vous plaise illuminer
L'obscure nuyct de ma triste pensée.   [I, p. 87]

Les yeux de nostre beau iour.   [I, p. 166]

Ce beau Teint, qui notre seiour
Embellist encor' d'un beau Iour.   [I, p. 172]

After the poetic possibilities of the term had once been recognized, a poetic world was created that was impregnated and illuminated by the idea of the day. The old symbolic antithesis of day and night, light and shade, which had pervaded ancient Christian religious verse, was revived. The idea of the day, therefore, became associated with spiritual, religious and ethical values; these associations superseded everything else linked with the concept of the day in the language of Corneille. The question of what the day saw, of what it might see and of what was in sympathy with it was implicitly present as a mental concept in all metaphorical uses of *jour*. Ronsard called the cockerel *aime-jour*:

> Debout doncq, aube sacrée,
>   Et recrée
> De ton beau front ce troupeau,
> Qui, pour toy, pend à la gaule
>   De ce saule,
> D'un coq aime-jour la peau.   [VI, p. 365]

The phrases *mettre au jour, voir le jour, montrer au jour* now appeared in French for the first time.

> Ronsard:
>   Le vers bien accomply,
> Qui tirant leurs noms de l'oubly,
> Plongez au plus profond de l'onde
> De Styx, les a remis au jour.   [II, p. 370]

> Ne laisse pourtant de mettre
> Tes vers au jour.   [II, p. 443]

>   Une femme en gesine,
> Qui avecque douleur met au jour ses enfans.   [V, p. 241]

>   Il est temps que tu (= luth doré) voises
> Donner plaisir aux oreilles Françoises
> Rompant l'obscur du paresseux séjour,
> Pour te monstrer aux rayons du beau jour.   [V, p. 283]

> Et ce present mettra ton beau renom au jour. [VI, p. 204]

> Du Bellay:
> Ce qui au jour se montre.   [I, p. 170]

> O que peu durable
> (Chose miserable)
> Est l'humaine vie!
> Qui sans voyr le Iour
> De ce cler Seiour
> Est souvent ravie.   [I, p. 184]

In order to provide an approximate idea of the manner in which the day of light, after a long period of linguistic barrenness, now bore leaves and flowers like the branches of a tree in spring, we represent this process in graphic form, setting out those terms which came to be associated with the word *jour*, in so far as the word related to the day as an experience, in the form of a tree, which may be visualized as gradually spreading its branches through the language. Our collection of terms which came to be closely linked with the word *jour* makes no pretensions to completeness.

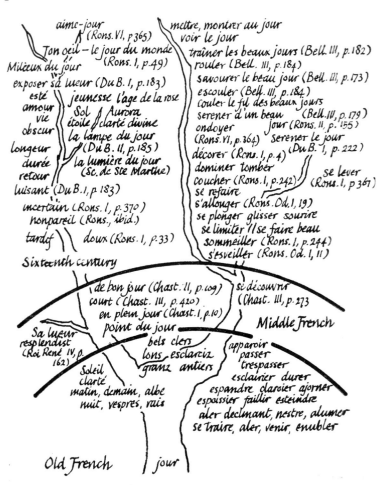

Notes

1 Heinrich Wölfflin, *Renaissance und Barock* (second edition), p. 22.
2 Eneas, v. 9935.
3 R. Glasser, '*Mi pare mill'anni*. Geschichte einer Redensart', *Roman. Forsch.*, 58, pp. 41–113.
4 *Œuvres* (ed. Pifteau), II, p. 213.
5 *Œuvres* (ed. Moland, new edition by H. Clouzot, Paris, Garnier), II, p. 374 (*Epistre à Jehan Bouchet*).
6 *Œuvres complètes* (ed. Blanchemain, Paris, 1857–67), I, p. 339.
7 *Ibid.*, I, p. 433.
8 *Œuvres Françaises* (ed. Marty–Laveaux, Paris, 1866–67), II, p. 185.
9 *Poésies*, p. 122.
10 II, p. 7 (Nov. 1).
11 *Op. cit.*, II, chap. 41.
12 Ed. Reichenberger, Tübingen, 1963.
13 *Œuvres*, I, p. 138.
14 Moland–Clouzot, II, p. 356.
15 *Les Tragédies de Montchrestien* (ed. Petit de Julleville, Paris, 1891).
16 *Œuvres*, II, p. 57. Man's power prematurely to dispose of the future, followed by disillusionment, is an ancient literary theme.
17 *Gargantua*, I, 33.
18 *Vita di Benvenuto Cellini, scritta da lui medesimo* (ed. Campori, Milan, 1873), II, chap. 16.
19 *Odes* (ed. Vaganay, Bibl. Romanica, II, p. 83).
20 Ariosto, *Orlando Furioso*, II, 17f.; Garcilaso de la Vega, *Égloga Segunda*, vv. 126f.; Ronsard, *Franciade*, 4; Shakespeare, *Macbeth*, IV, 1; Camões, *Os Lusiadas*, X; Voltaire, *Henriade*, VII.
21 *Œuvres Complètes* (ed. Vaganay), IV, p. 450.
22 Classiques Garnier, Paris, 1929, p. 124.
23 R. Garnier, *La Troade* (ed. Foerster, Heilbronn, 1882).
24 R. Garnier, *Antigone* (ed. Foerster, Heilbronn, 1883).
25 Montchrestien, *Tragédies*, p. 209.
26 Huizinga, *op. cit.*, pp. 38–75.
27 *Les Quatre Ages de l'Homme*, chs. 153–60.
28 *Ibid.*, ch. 159.
29 Alfred von Martin, *Mittelalterliche Welt- und Lebensanschauung im Spiegel der Schriften Coluccio Salutatis* (Munich, 1913), p. 41.
30 *Œuvres*, II, p. 184.
31 *Ibid.*, V, p. 368.
32 *Ibid.*, V, p. 370.
33 E. Picot, *Recueil général des Sotties* (Paris, 1902–12), II, p. 148.
34 *Gargantua*, I, 23.

35 *Gargantua*, IV, 3.
36 Agnolo Firenzuola, *Opere* (ed. Bianchi, Naples, 1864), I, p. 75.
37 Bibl. Romanica, p. 33.
38 *Op. cit.*, I, p. 71.
39 K. Vossler, *Frankreichs Kultur*, pp. 96–8.
40 A study of the age from the standpoint of the history of thought cannot dispense with an account of its conception of time. See Paul Meissner, *England im Zeitalter von Humanismus, Renaissance und Reformation* (Heidelberg, 1952), pp. 259–69.
41 Ed. Vaganay, Bibl. Rom.
42 V, pp. 181ff.
43 V, p. 414.
44 *Trionfo di Bacco ed Arianne*.
45 R. Belleau, *L'Heure; Œuvres*, I, pp. 59f.
46 Gauthier–Ferrières, *Anthologie des Écrivains des XVe et XVIe siècles* (Paris, Larousse), p. 135.
47 *Les Tragiques*, V: *Les Fers* (ed. Read, Paris), II, pp. 84–5.
48 Eugen Wolf, *Petrarca. Darstellung seines Lebensgefühls* (Berlin, 1926), p. 32.
49 Monstrelet, I, p. 343.
50 *Ibid.*, I, p. 5. Roi René, I, p. 56.
51 *Paradiso*, XV, v. 11.
52 *Deffence et Illustration de la Langue françoyse*, II, 2.
53 Ronsard, V, p. 256.
54 Ronsard, V, p. 326.
55 Ie fay tout, ie peux tout, ie lance ma parole, Comme vn foudre bruyant, de l'vn à l'autre pole. *M. Antoine*, vv. 1360–1 (ed. Foerster, Heilbronn, 1882).
56 Albert Thibaudet, 'Paysages', *Nouv. Revue Franç.*, July 1927, p. 87.
57 'El que desprecia el aplauso de la muchedumbre de hoy, es que busca sobrevivir en renovadas minorías durante generaciones— quiere prolongarse en tiempo más que en espacio.'—M. de Unamuno, *Del sentimiento trágico de la vida* (third edition), p. 60.
58 *Roman de la Rose*, vv. 361–91.
59 *Paradiso*, 10.
60 Belleau, II, p. 174.
61 Belleau, III, pp. 184, 197.
62 Belleau, III, p. 174; Passerat, *Recueil des Œuvres Poétiques* (Paris, 1606), p. 442. (Darmesteter–Hatzfeld, *Morceaux choisis . . . du XVIe siècle* (Paris, 1896), p. 274.)
63 Ronsard, *Odes*, I, Ode 10.
64 Jacob Grimm, 'Rede über das Alter', *Kleinere Schriften* (ed. Speidel, Berlin, 1911), p. 150.
65 Gauthier–Ferrières, *op. cit.*, p. 95.

66 *Deff. et Ill.*, I, ch. 9.
67 *Ibid.*, I, ch. 10.
68 Darmesteter–Hatzfeld, *op. cit.*, p. 193 (*La succession des Empires*).
69 *Les Antiquitez de Rome*, III (ed. Droz, Lille and Geneva, 1947).
70 *Œuvres complètes*, II, p. 225.
71 *Op. cit.*, chap. 69.
72 Cervantes, *Don Quijote*, I, 11 (ed. Marín, Madrid, 1933).
73 *Ibid.*, I, 15.
74 *Roman de la Rose*, vv. 17031f.
75 Alain Chartier, *Le Curial* (ed. Heuckenkamp, Halle, 1899), p. 27.
76 Bartsch–Wiese, *Chrestomathie de l'ancien français* (twelth edition, Leipzig, 1920), No. 89 (*d*).
77 Bartsch–Wiese, No. 94.
78 Jacob, *op. cit.*, p. 420.
79 *Poésies*, p. 96.
80 *Ibid.*, p. 183.
81 *De brevitate vitae*, 8.
82 Belleau, III, p. 184.
83 *Die vierundzwanzig Sonette der Louize Labé*, German trans. by R. M. Rilke (Leipzig, Insel Verlag), sonnet 14.
84 Rudolf Stadelmann, *Vom Geist des ausgehenden Mittelalters* (Halle, 1929), p. 115.
85 Che più d'un giorno è la vita mortale?
   Nubil' e brev' e freddo e pien di noia;
   Che pò bella parer, ma nulla vale.

   Trionfo del Tempo
86 See Friedrich Schürr, 'Ein Leitmotiv der Renaissancelyrik bei Ronsard und Malherbe', *Die Neueren Sprachen*, 40 (1932), pp. 278ff., in which Ausonius' *De Rosis* is adduced as a model for the new poetry of the rose.
87 M. Gorce, *Le Roman de la Rose* (Paris, 1933), p. 33.

5

The humanistic evaluation of time[1]

The first person to whom time appeared as a matter in its own right—one which affected him personally, and moreover with no reference to the beyond—was probably Petrarch. It was he who introduced the concept of 'the value of time' into contemporary literature:

Non solebat mihi tempus esse tam carum
[Lib. XVI, ep. xi (Fracassetti II, p. 397)]

Inaestimabile tempus est.   [*Ibid.*, II, 398]

Non æstimabam tempus suo pretio, scribebam amicis: imputabam laborem corporis animi taedium, pecuniae impensam: tempus ultimum erat. Nunc video: primum esse debuerat.   [*Ibid.*]

Among the arguments adduced by Petrarch in a letter to Charles IV with the purpose of arousing him to travel to Rome and set on foot the restoration of the Roman Empire, reference to the preciousness of time held a prominent place:

Pretiosissima, imo vero inextimabilis res est tempus, et cuius solus 'avaritiam' doctorum hominum commendat auctoritas. Pelle moras igitur, et quod grande aliquid aggressis utilissimum est, singulos dies magni extima! Ea te cogitatio parcum temporis efficiet, ea te coget ut venias et inter adversitatum nostrarum nubila speratum nobis augustissimae tuae frontis lumen ostendas.[2]

Petrarch named the writer under whose influence he used the concept and argument of the value of time. It was Seneca. In all such utterances the humanists showed themselves to be influenced by the pronouncements of the ancient philosopher on the subject of time. No one gave precious time such a predominant place in his consciousness as did Seneca. Since the age of Hesiod, time had been the object of philosophical contemplation.[3] Seneca has no place in the history of philosophy, which concerns itself with the nature of time, by virtue of the fact that he viewed time exclusively from a practical standpoint. It was for him an essential feature in the conduct of life; indeed, in no ethical system was it of such overriding importance as with him. What is remarkable about his pronouncements on the subject of time is the fact that he linked it closely with his conception of the individual, so much so indeed that he made the person and time coalesce. The injunction 'vindica te tibi'[4] is nothing more than an admonition to utilize time on one's own account, for time was a valuable possession.

> Quem mihi dabis, qui aliquod pretium tempori ponat, qui diem aestimet, qui intelligat se quotidie mori?

In contrast to all other things, time is something we truly possess within ourselves:

> Omnia, Lucili, aliena sunt, tempus tantum nostrum est.

It was, however, treated with downright barbaric unconcern. People did not know how to make anything of it. To utilize it aright and to fill it meaningfully was a most desirable goal, and one which called for the highest degree of wisdom. To this end one must also be master of time, 'temporis dominus'[5]—that is to say, one must not be hindered either externally or internally in shaping it.

Seneca answered the question of the ideal content with which to fill time in these terms:

> Soli etiam otiosi sunt qui sapientiae vacant, soli vivunt.
>
> [*De brev. vitae*, XIV]

For one capable of turning it to account, its limits extend far into the past and even into the future. He is at home in the *aevum*. The striving to overcome temporal contingency and the moment, to break the trammels of time, and to attain to that state called timelessness, found an eloquent advocate in Seneca. His temporal appetite ranged into the past and the future:

> Transiit tempus aliquod: hoc recordatione comprehendit; instat: hoc utitur; venturum est: hoc praecipit. Longam illi vitam facit omnium temporum in unum collatio.   [XV]

The considerations of the Stoic philosopher on the question of time have to do only with the individual and his personal concerns, not with the community or the world.

The injunction to protect what is truly one's own from loss and to cultivate it is closely related to the Christian evaluation of the individual human soul. The awareness of one's own spiritual life, championed by Christianity, and of the particular destiny of the individual, is seen to be akin to the urgency with which Seneca advocated the proper possession and enjoyment of time. The cast of mind and mood conducive to the wise economy of time belong to old age rather than to youth. It is a matter of protection and retention rather than of acquisition. The kinship between time and material possessions (money), as set forth by modern Anglo-Saxon thought, found an early counterpart in *De brevitate vitae*, in so far as time and property were related in it. But in what a relationship!

> Nemo invenitur, qui pecuniam suam dividere velit: vitam unusquisque quam multis distribuit!   [III]

Here time does not stand in a dependent relationship to external things, but is set as something unique, as a value in itself, above all possessions, so that the latter may be evaluated by the degree of freedom they grant to their owner in regulating his time.

From these reflections on the most important problem life presents to man, the proper treatment of time, there necessarily arises a particular assessment of human intercourse. It is evaluated *sub specie temporis*; time appears as a commodity men request from one another, one which they mutually give and take, steal and grant. This conception, a commonplace with us, was unknown to the Middle Ages. It occupied an important place in Seneca's thought:

> Cum video aliquos tempus petentes et eos, qui rogantur, facillimos.   [VIII]

> Dati nobis temporis spatia.   [I]

> Duc, quantum ex isto tempore creditor, quantum amica, quantum rex, quantum cliens abstulerit, quantum lis uxoria, *etc.*   [III]

> Cum interim fortasse ille ipsi qui aliquoi vel homini vel rei donatur dies ultimus sit.   [III]

Dicere solent eis, quos validissime diligunt, paratos se partem annorum suorum dare.   [VIII]

Nihil largitione detrahitur.   [XI]

Horum nemo annos tuos conteret, suos tibi contribuet.   [XV]

Seneca's place in the history of the consciousness of time may be compared with the merit of those painters who for the first time perceived, and attempted to represent, the presence of the air and atmosphere as a medium existing between things. In this way he, and those influenced by him, became aware of the presence of time among men—not time as a general order, with its objective measurable relationships (*le temps social*), but as a personal possession that may be given away.

The repetition and variation of this theme by the humanists, as we shall see, are proof in themselves of the wide-ranging influence exercised during the centuries of the Renaissance by Seneca's reflections on time. They probably had their effect in tempering barbaric notions in the matter of dealing with one's own time and that of others, and—in the case of the most amenable minds, at least—in strengthening a new awareness of the part played by time. By virtue of this influence, what was said in the letters to Lucilius and in the treatise *De brevitate vitae* had a much more decisive effect on the modern view of time than had all theoretico-philosophical speculations as to its true nature.

The subjective element of this effect was in any case made possible by a novel view of the essence and value (a view already set forth by antiquity) of the individual, and by a new conception of *dignitas hominis*. Where the human being is of no value, time is worthless also. The new orientation in the world of values in connection with the elevation of time to the throne went far beyond the orthodox influence exerted by Seneca and the Stoics on the age of humanism. The value of time and allied considerations were, at least in circles representing a consciously humanistic point of view, a general civilizing force, as Seneca's reflections ('une espèce de menue monnaie de la sagesse'[6]) show. This newly gained insight brought a sense of obligation more than it inspired enthusiasm or emotion. The poet speaks of the passing of time, its melody, the melancholy of transitoriness, the fresh joy of *carpe diem*; the moralist of the value of time. Petrarch tells of his experience of time in the *Canzioniere* and the *Trionfi*, and of his evaluation of it in the Epistles. Renaissance poetry had nothing to say of the evaluation of time.

To heed the value of time is consistent with a mature mental attitude. Like the youth who believes his life-time to be spread out before him like a boundless ocean, a person will thoughtlessly fritter away hours and days if a task is not incumbent on him which admonishes him to make a duty of economy. An individual for whom the future begins to be short and uncertain will, with his wisdom of the art of living, make a well husbanded possession of his time, as the aged Montaigne did. If the humanist advocates thriftiness with time, he does so from an awareness of his tasks and mission rather than from a feeling of old age. In addition, the world at the close of the Middle Ages had become vast and rich; the fullness of those things placed before him in the present and the past in the expanded universe, and the more rapid rhythm of life, made him realize that the short span of time allotted him was insufficient for dealing with these problems, to which this new wealth had added.

The idea of the value of time was a new piece of wisdom for the Middle Ages, one which was, therefore, proclaimed with an exhaustiveness befitting it, even initially. Chaucer's landlord of the Tabard inn admonished the host of pilgrims to make proper use of time, pointing out its value inherent in its irrevocableness, a value which had been recognized by Seneca:

> Wel kan Senec and many a philosophre
> Biwaillen tyme moore than gold in cofre;
> 'ffor losse of catel may recouered be,
> But losse of tyme shendeth us', quod he,
> 'It wol nat come agayn, with-outen drede,
> Namoore than wol Malkynes maydenhede,
> When she hath lost it in hir wantownnesse!
> Let us nat mowlen thus in ydelnesse!'
> [*Canterbury Tales*, vv. 4445–2]

In relation to the story in which they occur, the only purpose of these words is to induce the Man of Lawe to recount his tale. From the aesthetic point of view, however, they have to do with the satisfaction Chaucer takes in establishing the time of day or the season[7] and thereby creating a temporal link in his mind with the reflection bound up with it.

At the close of the Middle Ages the opportunity to comment on time was always eagerly seized upon. As far as the above quotation is concerned, it seems to be a question of communicating an edifying piece of wisdom rather than Chaucer's own experience of life. For Chaucer the value of time was still not axiomatic, and the equation 'time = valuable possession' was still regarded and expounded as a comparison, because it was unusual. Almost two centuries later we read in the *Heptameron* of Marguerite de Navarre:

> Et tant plus le temps de leur propos estoit abbregé par contraincte, et plus leurs parolles estoient dictes par grande affection; car ilz desroboient le temps, comme faict ung larron une chose precieuse.
> [III, 1]

In this instance also the speaker did not talk, as the authors of the succeeding classical age were to do, of 'valuable time' but is still aware of the figurative association binding the words 'time' and 'valuable'. Marguerite did not write something like this: 'They did not steal time, so valuable for them', but 'They stole time, as a thief steals something precious'. This proves that the image was still fresh and had not become a cliché. The *chose precieuse* corresponded exactly to the *preziosa cosa* of which Bandello spoke:

> Oh con quanta diligenza, fatica et amore attenderebbono a farsi disciplinati, con quanta cura dispensarebbero l'ore, acciò che così preziosa cosa come è il tempo, che è irreparabile, non si spendesse vanamente, non si gettasse via, non si consumasse in cose frivole e di nessun momento![8]

The association between the thought of learned activity and the value of time was so close that, when it was a question of literary pleasures and the cultivation of things of the mind, one inevitably spoke of that precious thing, time. The notion that time represented a value was thus, in its first appearance in modern thought, confined to a quite definite sphere. This was the narrow humanistic view of life and evaluation of things. He who spoke of the value of time thought in that age neither of money, nor of progress in the struggle of life, nor of anything other than the humanistic mode of employing time for literary ends or those of general culture.

The realization of the preciousness of time led to the injunction —made, for example, by Seneca—not to waste it. It was a matter of employing it as meaningfully as possible. Here also Petrarch had the first word:

> Ut, quantum possibile est, brevissimum hoc tempus sine super-vacuis et inanibus curis agas.   [Lib. III, ep. viii (Frac. I, p. 155)]

He gave the following advice to a young ruler:

> Sit sane tenax famae propriae, sit parcus honoris, sit avarus tem-poris, sit largus pecuniae.   [Lib. XII, ep. ii (Frac. II, p. 165)]

He himself showed his desire for time as tending in two directions: on the one hand in his striving to have as much time as possible for himself, for his own unrestricted use; on the other, in his wish to wrest something from time even after his death in the shape of posthumous fame. Both these expressions of a desire for time sprang from a single spiritual root: concern for the broadening and unfolding of the scope of his own life in opposition to forces he felt to be hostile. Petrarch knew how to rid his life of obligations of all kinds, which by making inroads into his time would narrow the range of his own life.[9] He was the first to sense that time was part of one's own being and that in giving of his time he was giving of himself.

For him, time was a disquieting force. In the *Trionfi* it has the final say in earthly matters; it triumphs over love, chastity, death and fame. Only the eternity of God still stands over it. Here occurs the form of thought that became a pervading motif in Renaissance literature: time against love, time against fame.[10] With the word *tempus* Petrarch designated, not so much an order and gradation of values, but rather the most important fact of his experience of life. The latter was shot through with melancholy inspired by the brevity of the course of the world. The idea of *brevitas*, applied to all manner of phenomena, constantly recurs in his writings. His rhythmical sense for the tempo of happenings, for their incomprehensible and irresistible gliding and flowing, was sharpened.[11] In this respect the poetry of the Renaissance was altogether under his influence. The most important change, however, which the idea of time underwent in

Petrarch's case was the weight and significance he attached to it amongst the host of things that engaged his attention. Preoccupation with time was a distinguishing feature of the dying Middle Ages. Its appearance in personal guise in the French allegories was a phenomenon comparable to its symbolization in the *Trionfi*. It presented itself to consciousness as something resembling a personal being, as an independent force of great influence.

If anything might be regarded as the enemy of the humanist, against which he was ever struggling, then it was the destructive effect of time, which threatened his life and work. His ultimate goal was the overcoming of transitoriness. For him, time was like a valuable and exalted power of fate which imparted greatness to those things able to pit themselves against it. The newly enhanced love of antiquity was nourished by the feeling that the centuries which had elapsed between the time of Augustus and that of Petrarch had been a test for the values and writings of antiquity—one which they had successfully withstood. The most living symbol of worth proved by time was, for the humanists, the book. It was a better pledge of lastingness than any edifice. Rome had fallen in ruins, but its intellectual riches lived on in its writings.

> Le corps de Rome en cendre est devallé,
> Et son esprit rejoindre s'est allé
> Au grand esprit de ceste masse ronde.
> Mais ses escripts, qui son loz le plus beau
> Malgré le temps arrachent du tumbeau,
> Font son idole errer parmy le monde.[12]

The collecting of books and the search for manuscripts is one form of a longing to possess that which withstands time. Books make it possible to 'catch up' with time. They have themselves resisted the centuries and should also point out to the young author the means of overcoming them. This gives rise to that small-scale hunger for time which sees the chances of a successful struggle against it reduced with every wasted minute. The humanist is fully aware that he must bring in the harvest which will assure his future—and with it the perpetuation of his name. This is the feeling which a more recent poet, John Keats, expressed in the lines shown overleaf.

When I have fears that I may cease to be
Before my pen has glean'd my teeming brain,
Before high-piled books, in charactr'y,
Hold like rich garners the full ripen'd grain.

It seems to be in the nature of things that nothing should so peremptorily demand time and its unrestricted use as the learned and literary life. Nothing claims the attention of humanity as it does. No purposes seem to be so closely bound to the condition of having time as humanistic ones. He who gives the humanist money and takes away his time serves him ill. For the humanist time is not exchangeable and alienable, for he has recognized it as distinct from other values. The desire for leisure is the very heartbeat of humanism. He who seeks to deny the humanist his desire deals him a sure and deadly blow by robbing him of his time. No deprivation is harder to bear than that of leisure. Work, exertion, disease and poverty seem deplorable chiefly because they interfere with the free deployment of time.

Piget me non laboris, sed temporis. Nihil apud me preciosius tempore, omne id temporis periisse mihi puto, ubi non de melioribus litteris aliquid imbiberim.[13]

This attitude to life, consisting in a desire to have time at one's disposal, meant that the future was not a matter of indifference. Whereas the Renaissance often advocated at once a subjective unconcern for what was to come and a vigorous affirmation of the present, the true humanist felt his pleasure in things of the day to be purest and strongest when he believed he was working for the future. He never lived entirely in the present. *Otium*, which held a prominent place in Petrarch's thought, was a condition for that which was to endure for all time. Without leisure there could be no posthumous fame. If the French poets of Ronsard's circle showed a predilection for the word *loisir*, they betrayed the impact of humanism.

The exalted status attained by time is proved by the circumstance that Leon Battista Alberti accorded it a truly singular place amongst human values. He allied it with the body and the soul to form the trinity of those things which can be described as really our own. Time becomes our possession by virtue of our utilizing it, and that to the ends of the specifically humanistic ideal of life:

S'egli è chi l'adoperi [il tempo] in lavarsi il sucidume et fango quale a noi tiene l'ingegno et lo intelletto inmundo, quale sono l'ignoranzia et le laide volonptà et brutti appetiti, et adoperi il tempo in imparare, pensare et exercitare cose lodevoli, costui fa il tempo essere suo proprio; et chi lascia trascorrere l'una ora dopo l'altra otiosa sanza alcuno onesto exercitio, costui certo lo perde.[14]

The eagerness with which Alberti strove to reach the goal of making time truly his is well known. For him, time was something costly beyond measure:

. . . il tempo quanti e a' beni del corpo et alla felicità dell'anima sia necessario, voi stessi potete ripensarvi, et troverete il tempo essere cosa molto pretiosissima.[15]

Adopero tempo quanto più posso in exercitii lodati, non l'adopero in cose vili, non spendo più tempo alle cose che ivi si richiegga a farle bene. Et per non perdere di cosa sì pretiosa punto, io pongo in me questa regola: mai mi lascio stare in otio, fugo il sonno, né giacio se non vinto dalla strachezza, che sozza cosa mi pare senza ripugnare cadere et giacere vinto.[16]

With the idea of time he associated the thought of realizing a lofty ideal of life. His words are evidence of an insatiable hunger for time, which saw every wasted minute as a loss that can never be made good. Thus the first requirement of his philosophy of life was to let no time remain unused:

Per questo, figliuoli miei, si vuole observare il tempo, et secondo il tempo distribuire le cose, darsi alle faccende, mai perdere una ora di tempo. Potrei dirvi quanto sia pretiosa cosa il tempo, ma altrove sia da dirne con più elimata eloquentia, con più copia di doctrina che la mia. Solo vi ricordo a non perder tempo . . .[17]
. . . stima ogni tempo essere perduto se non quello el quale tu adoperi in virtù.[18]

Michelangelo also used like turns of expression:

non ritrovo
Fra tanti un giorno che sia stato mio.   [Madrigale 52[19]]

He formulated the value of time in negative terms:

Che non è danno pari al tempo perso.   [Madrigale 51]

The poems of Michelangelo reveal how men's minds were now pondering on the matter of time. For him it was more than a

central concept; it was a lasting source of disquiet to his consciousness, something melancholy, which lent a gloomy hue to everything he said. For him, reflection on the subject of time went far
beyond the function of cultural apparatus and of incidental remembrance. One can almost speak of an obsession with time,
which diverted his attention from all humanistic and other
contemporary preoccupations.

In those centuries time as a possession was spoken of with
varying significance and from differing standpoints. With Leon
Battista time, if meaningfully deployed, counted as *tempo nostro*.
In the French of the later Middle Ages *avoir temps* was a con-
dition for success in the rough and tumble of human relations.
Rabelais in his turn spoke of his 'hours'[20] and Shakespeare's
Macbeth said:

> Let every man be master of his time
> Till seven at night.   [III, 1]

The image of the person who has time at his disposal as a possession
thus became plastic in the language. The unrestricted mastery
over time was set up by Rabelais in the classical manner as a new
ideal.

If, however, we examine the ideals of the deployment of time
more closely we may perhaps be able to draw a dividing line
between humanism and the Renaissance. That which we recognize as the temporal avarice of the humanists contradicts our
notion of the Renaissance individual, who deployed his time in a
liberal and carefree manner. The banishment of clocks from the
abbey of Thélème provides an instance of this attitude. Rabelais
himself, however, exemplified the temporal avarice of the
humanists, with his desire to see the young giant brought up to
utilize time rigorously.[21] The two basic attitudes—free enjoyment of time, on the one hand, and its thrifty husbanding on
the other—cannot always be distinguished if one goes by individuals. People contradicted one another in their conduct of
life and their thought. There are, nevertheless, fundamental
expressions of opinion which remove all doubt as to which attitude
is the more inward and profound and must be inculcated in a
person.

What Erasmus said on this subject is relevant here. According to him, the essence of the deployment of time consisted in profitable study. In a letter of 1479 to Christian Northoff he gives us an idea of his conception of a well used day:

> Nocturnas lucubrationes atque intempestiva studia fugito; nam et ingenium extinguunt et valetudinem vehementer offendunt. Aurora Musis amica est, apta studiis. Pransus aut lude, aut deambula, aut hilarius confabulare. Quid quod inter ista quoque studiis locus esse potest? Cibi non quantum libidini, sed quantum valetudini satis est sumito. Sub coenam paulisper inambula, coenatus idem facito. Sub somnum exquisiti quippiam ac dignum memoria legito, de eo cogitantem sopor opprimat, id experrectus a teipso reposcas.[22]

This is a rationing of the time at one's disposal whereby one does not act merely according to the principle of humanistic temporal avarice, but whereby time appears as a manifold organism with whose particular laws one must be familiar if one is to use it aright. Accordingly, Erasmus indicates those times particularly favourable to study. He warns against an excess of it, and if one thinks of the radicalism to which the young Gargantua was bound by his tutor in the utilization of time, this warning seems understandable. The antique admonition to use time aright came straight from the humanist heart.

> Plinianum illud semper animo insidiat tuo, omne perire tempus quod studio non impertias.   [Plin. Ep. 3, 5, 16[22]]

Seneca and Pliny were not the only figures to be recognized as models for this attitude. The example of Cato was pointed to:

> Is enim, quicquid temporis a lectione vacuum labi pateretur, cum in Senatu accederet, libellum quempiam secum deferebat cuius particulam aliquam, interea dum Senatus colligebatur, lectitabat.[23]

The thing and the word, the attitude and the formulation, were imitated from antiquity. But the zeal and assiduity with which this evaluation of time was interpreted in words and deed were new. To waste no time is seen here above all as constituting the ethical duty and purposeful attitude of the superior and noble human being. To waste time was a form of barbarism. This view also lay at the root of the terms with which Erasmus related the coming into being of his *In praise of folly*:

> Superioribus diebus cum me ex Italia in Angliam reciperem, ne totum hoc tempus, quo equo fuit insidendum, ἀμούσοις et illiteratis

fabulis tereretur, malui mecum aliquoties vel de communibus studiis nostris aliquid agitare, vel amicorum (quos hic ut doctissimos, ita et suavissimos reliqueram) recordatione frui.

We find the same justification for the creation of a literary work coming from the pen of the same author in similar circumstances and in connection with another matter:

Nuper igitur cum itineris labore delassatis equis dieculas aliquot apud divum Audomarum cogerer subsidere, ne tempus hoc omnino periret studiis, de parando tibi xeniolo coepi cogitare.[24]

The ideal of a life conducted on these lines would appear to be the complete filling up of existence by literary work. What does not serve the pursuit of the learned sciences is a waste of time. We know that Erasmus evaded obligations in order to remain master of his time. He justified his refusal of a prebend in the following terms:

Ego huic ocio meo et studiorum laboribus omnia mea posthabeo.[25]

Agricola expressed himself in like manner:

Am I to assume direction of a school? How then shall I find time for my studies and for creative scientific activity on my own account? Where shall I find one or two hours to interpret an author, since attending to the boys takes up most of the time, so much so that the teacher's patience is exhausted by the fact that he cannot devote his free time to his own learning, but requires all of it to rally his strength?[26]

In 1536 the Belgian humanist Nicolas Clénard requested in a letter to Hieronymus Aleander, Archbishop of Brindisi and Cardinal, that he might be relieved of the duty of the daily reading of the breviary for the sake of his studies:

nec te clam est, optime praesul, quanta semper temporis cupidine ducamur οἱ τοῖς βιβλίοις ἐπικεκυφότες καὶ ὡς σχολῆς μᾶλλον εὐπορειν βουλόμεθα ἢ βασιλικῶς πλουτεῖν.[27]

Time is something so important that it plays a part in many decisions one has to take; indeed, on occasion it dictates the overriding argument for adopting a particular course. By the same token, the aversion of many humanists to marriage sprang from a fear that domesticity might make undue demands on one's time. The married Thomas More had greater difficulty than his single friends in finding time for himself, that is, for his literary pursuits.

> Dum foris totum fermè diem aliis impertior, reliquum meis, relinquo mihi, hoc est literis, nihil.[28]

He tells us in graphic terms how his duties as State official and head of a house and family encroached on his time. 'Quando ergo scribimus?' He goes on to speak of sleeping and eating, which take up a good deal of time. He wrested his leisure from both of these activities:

> At mihi hoc solum tempus acquiro, quod somno ciboque suffuror.[28]

Another complained that

> A natali Christi in hodiernum usque diem ne horula quidem data est, quam literis impendissem.[29]

These are a few examples of the many grievances aired in those centuries. There were probably never so many complaints about lack of time as there were then. It is a remarkable fact that much that was said about the significance of time did not speak of its actual use, but mainly took the form of expressions of intention, of strivings, wishes or laments, from which we may see how little the humanists' imaginings of an ideal deployment of time were founded on reality. We may often detect the inner strength which imparted to them a truly heroic attitude when they wished to take a stand against the basic elements of their environment which robbed them of their time. Within the limits which communal life imposes on the individual, the humanist may make his time as precious as he chooses. From this attitude there arises a mode of life which consists in not putting one's time at the disposal of anybody and everybody, for not every social contact is (in the literal sense) 'worthwhile'.

In a letter to Lascaris of 12 June 1520 Guillaume Budé wrote the following sentence, available to us only in translation:

> Je pense, en effet, qu'il est nullement convenable à ceux qui s'occupent d'études littéraires d'aller perdre ainsi leur temps avec des ignorants.[30]

Seneca had already adumbrated this weighing up on a balance which set time and intellectual gain against one another:

H

> Cum me amicis dedi, non tamen mihi abduco, nec cum illis moror,
> quibus me tempus aliquod congregavit aut causa ex officio nata
> civis, sed cum optimo quoque sum.   [*Ep. ad Lucilium*, 62]

The deployment of time must not be decided by chance. This
new responsibility necessitated the making of more clear-cut
choices in order to make something of time. It passes so quickly,
it is nothing, and yet everything is to be expected of it.

The evaluation of time as a means to the satisfaction of a restless
eagerness for learning, of an impulse to gain culture, perfection
and immortality, is one of the distinguishing marks which make
it possible to recognize those humanists truly worthy of the name.

> Dum non habent quod agant, litteris bonis non desinunt incumbere,
> et ne tempus frustra labatur, quotidie eruditiores evadere conten-
> dunt.[31]

may be taken as a general and characteristic opinion. Whoever
was heedless of his own time was no humanist. That which imparted its particular aspect to historical humanism at the beginning of the modern age was less the individual cultural objects
with which humanism concerned itself than a particular evaluation, which had followed the promptings of a newly awakened
and powerful instinct, of the elementary circumstances of life,
of which time was one.

The right use of time had become a problem that could not
be evaded unless one was confined by monastery walls and had
thus renounced the opportunity of deploying it as one chose. If
we now often come across the phrase *employer le temps*, which
had already become current in Old French,[32] we may see it as
reflecting something of the preoccupation of the age with the
shaping of the day and of life. On the lips of the humanists it
received the connotation of a decisive choice between activities
of different values. Du Bellay contrasted the nobler deployment
of time with the baser:

> Les allechementz de Venus, la gueule et les ocieuses plumes ont
> chassé d'entre les hommes tout desir de l'immortalité: mais encores
> est ce chose plus indigne, que ceux qui d'ignorance et toutes especes
> de vices font leur plus grande gloire, se moquent de ceux qui en
> ce louable labeur poëtique employent les heures que les autres
> consument aux jeux, aux baings, aux banquez, et autres telz menus
> plaisirs.   [*Deff. et Ill.*, II, 5]

And likewise:

> Reçoy donques ce present, tel qu'il est, de la mesme volonté
> que je te le présente: employant les mesmes heures à la lecture
> d'iceluy, que celles que j'ay employées à la composition: c'est le
> temps qu'on donne ordinairement au jeu, aux spectacles, aux
> banquetz, et autres telles voluptez de plus grans fraiz, et bien
> souvent de moindre plaisir.[33]
> Nous emploierons le temps à la cognoissance des sciences et de la
> philosophie, lequel ils estoient contraints d'emploier à la cognoissance
> des langues.[34]
> ...N'est-ce pas grand pitié que deux si grands personnages, au lieu
> d'employer le temps à des escrits qui les pouvoyent rendre admi-
> rables, l'ayent employé à des disputes touchant leur langage ma-
> ternel.[35] L'Empereur Severe, qui emploioit tout son temps à la
> Chasse, finit miserablement ses jours.[36]
> Or sus, pour voir, faictes vous dire un jour les amusemens et
> imaginations que celuy là met en sa teste et pour lesquelles il
> destourne sa pensée d'un bon repas et plainct l'heure qu'il emploie
> à se nourrir, vous trouverez qu'il n'y a rien de si fade en tous les
> mets de vostre table que ce bel entretien de son ame.[37]

From the enhanced appreciation of time and from an awareness,
to which the Middle Ages were blind, of those things which went
to make the personal property of humanity there also developed
an attention and respect for the time of others. Regardless of
whether the many expressions of politeness and kindness we
encounter in the letters of the humanists were personally sincere
and genuine, they point to an ideal of communication between
men who strove to demonstrate their own worth and an exalted
conception of human dignity by means of a civilized manner of
dealing with their opposite numbers. Behind all that was said
there stood an image of the ideal human being.

Where there exists a fine feeling for personal considerations,
one readily senses whether one has overstepped their proper
limits. A person encroaches on the province of another by his
action and realizes that he has taken something away from him
in doing so. One who writes a letter to another is *ipso facto* in
his debt, for he at least requires him to devote some of his time
to reading it. Petrarch insisted that his friend should pay undivided attention to his letters and hence sacrifice some of his
time to him.[38] It is not surprising that the idea of giving time
should have played a part in the social life of the humanists.

That which for a coarser turn of mind and in less cultivated circles found no expression in language was here recognized as a gift. In general, the Middle Ages were unappreciative of the gift offered in virtue of one person sacrificing his time to another. The humanists, however, readily represented the little attentions people paid to one another, the pains they mutually took and the kindnesses of one sort or another which they exchanged, as the giving of time, with an assurance born of the fact that the sacrifice of time appeared as the highest that could be made. Recognition of a sacrifice made by another is in addition a sign of esteem. The great man who gives of his time gives of his own valuable life in so doing.

We find the earliest expressions of polite solicitude for another's time in the letters of the humanists, and particularly whenever it is a question of not burdening the highly placed and respected recipient with demands on his time. In a letter of Petrarch to Robert of Sicily we read:

> Sed iam metuo ne prolixitas in fastidium vergat; elegantissima quoque brevitas tua [the epigram coined by the king] ne longius vager admonuit.[39]

Erasmus wrote to John Colet in 1504:

> maiorem in modum te oro obsecroque ut posthac tantillum ocii suffureris studiis negociisque tuis, quo me nonnunquam literis tuis compelles.[40]

Again, on 31 October 1513 Erasmus wrote to J. Colet in these terms:

> Palam admonuit me tuus Gulielmus te literis scribendis occupatissimum esse, ne quid obturbarem.[41]

And Rabelais wrote thus to Maistre Antoyne Hullet:

> Ergo veni, Domine, et noli tardare, j'entends salvis salvendis, id est, hoc est, sans vous incommoder ne distrayre de vos affayres plus urgens.[42]

Such expressions show a courtesy differing from other, especially medieval, forms in that it included time in its considerateness. In contrast to this, a more external and ostentatious politeness took up the time of those persons who were the object of respect and deference, by reason of its diffuseness. Such a form is inferior to that of the humanists and is consequently much easier to practise and more often met with. Erasmus reported of Thomas More that 'he considered it effeminate and unworthy of a man to waste time with such foolishness'.[43]

Regard for the time of another is particularly called for if he be an important intellectual figure or scholar who knows how to put his time to good use. The sacrifice of time he is required to make then weighs even more heavily. Princess Elizabeth concluded a letter of 29 November 1646 to Descartes with the words:

> Je me laisse aller ici au plaisir de vous entretenir, sans songer que je ne puis sans pécher contre le genre humain, travailler à vous faire perdre le temps que vous employez pour son utilité, en la lecture des fadaises de votre affectionnée amie à vous servir.[44]

This consideration led to the development of a new, temporally formulated species of self-effacement.

The fact that time can be thought of as a gift and evaluated as such found expression in the language of the Renaissance. The phrase 'to give time' now became current in a new sense. It had, indeed, already existed earlier with the purpose of expressing the fact that one was granting another a period of time or setting a limit to one.

Chrestien:
Jusqu'a la feste saint Jehan
Te dona ele de respit.   [*Yvain*, vv. 2750–1[45]]

Commynes:
Ne leur donnèrent que troys heures pour se confesser et penser à leurs affaires.   [V, 17 (II, p. 202)]

Castiglione:
E darassi tempo al Conte di pensar ciò ch'egli s'abbia a dire.
[*Il Cortegiano*, I, 13[46]]

Du Bellay:
Il nous a faict humblement supplier et requerir que luy donnons & voulions octroyer le temps & delay de troys ans.[47]

Such turns became popular in seventeenth-century French:

Retz:
Il marcha vers Paris, en dessein d'arriver à Charenton, d'y passer la Marne et de prendre un poste dans lequel il ne pourrait pas être attaqué. M. de Turenne ne lui en donna pas le temps.[48]

La Rochefoucauld:
Il faut leur donner le temps de se faire entendre.

[*Réflexions diverses*, V[49]]

Racine:
Profitez du moment que mon amour vous donne.

[*Mithridate*, IV, 4]

Mme de Sévigné:
Ah, mon Dieu! Donnez-moi un peu de temps.

[Letter of 26 July 1691]

Chalussay:
Donnez-moy, s'il vous plaist, le temps de respirer.

[*Élomire Hypochondre*, V, 3[50]]

Perrault:
Mais elle demanda qu'avant que de paraistre
Devant son seigneur et son maistre
On luy donnast le temps de prendre un autre habit.

[*Peau d'Asne*]

Lesage:
A peine me donnèrent-ils le temps de descendre de cheval.

[*Gil Blas*, I, 8]

Je ne donnai pas a Leonarde le temps d'en dire davantage.

[*Ibid.*, I, 10]

Le bonhomme ne lui donna pas le temps de finir son discours.

[*Ibid.*, V, 1]

This *donner le temps* is the perpetuation of an ancient linguistic usage which originally had nothing to do with the humanistic conception of time. Now the expression was used in a new sense as opposed to its former one. It carried the implication of 'dedicating, sacrificing, giving a portion of one's own valuable time' to another (or to a thing), so that the time given was not a period but a real gift, deriving from one's own time. It often expressed a personal relationship. In this sense time was given primarily to a friend or other person to whom one felt indebted, and sometimes also to a thing deserving of the sacrifice. The following examples illustrate this turn of thought.

Boccaccio:
Questo uccello tutto il tempo da dover esser prestato dagli uomini al piacer delle donne lungamente m'ha tolto.   [*Decameron*, VII, 9]

Petrarch:
Et unam, precor, horam tuam relegendis unius diei mei actibus tribue!   [IV (Frac. I, p. 197)]

Du Bellay:
Pecheroy'-je pas (comme dit le Pindare latin) contre le bien publicq',
si par longues paroles j'empeschoy' le tens que tu donnes au service
de ton prince, au profit de la patrie, et à l'accroissement de ton
immortelle renommée? *[Deff. et Ill.,* dedication[51]]

M. de Navarre:
Si tous les matins vous voulez donner une heure à la lecture.
[Hept., prologue]

Henri Estienne:
A ceux qui me diront, qu'il faudroit avoir mangé beaucoup du
pain d'Italie, premier que pouvoir disputer si avant de son langage,
et que ce seroit le vray moyen d'en avoir telle congnoissance que
requiert mon entreprise, je respondray qu'ayant donné trois ans
de ma jeunesse à l'Italie, j'ay eu nonmoins le loisir que la com-
modité d'apprendre son langage.
[De la Precellence du Langage françois, pp. 191–2]

Estienne Pasquier:
Autant m'en est-il advenu voulant donner quelques heures à la
lecture de vos partisans. [Letter to P. Ramus, 1572[52]]

Olivier de Serres:
De tels utiles plaisirs, jouïra le Gentil-homme des champs, sans
destrac de ses affaires, s'il se mesure ainsi qu'il appartient, à ce que
ne s'abandonnant à ses plaisirs, il donne à la Chasse quelques heures
de son loisir. *[Le Théâtre d'agriculture,* p. 112]

Théophile de Viau:
Je sçay bien qu'un jour je me repentiray de ce loisir que je devois
donner à quelque chose de meilleur.[53]

Descartes:
M. Freinsheim a fait trouver bon à sa majesté que je n'aille jamais
au château qu'aux heures qu'il lui plaira de me donner pour avoir
l'honneur de lui parler.[54]

Mme de Sévigné:
Si vous lui donnez un peu de votre temps pour causer avec elle.
[Letter of 15 January 1690]

Gourville, qui n'a pas souvent du temps a donner.
[Letter of 7 October 1676]

Sa majesté, qui donne à madame la dauphine le temps qu'il
donnoit à Madame de Montespan. [Letter of 20 March 1680]

Shakespeare:
There is money; spend it; spend it; spend all I have, only give me

as much of your time in exchange of it, as to lay an amiable siege to the honesty of this Ford's wife.   [*Merry Wives*, II, 2]

The use of the expression 'to give time', as is apparent from these examples, was foreign to the Middle Ages, and moreover probably entered the language only under the stronger influence of antique modes of thought. This influence, exercised in particular by Seneca, is all the more understandable given that the idea of granting time unconstrainedly took its place in the temporal world of the imagination as it was handed on: time was the gift of a costly possession that must be carefully husbanded.

In the opinions we have so far considered, wishes and judgments are dealt with which are confined to the private and personal sector. The humanist first and foremost desired the opportunity of deploying time freely, considering it his primary concern to maintain his position in the face of forces and events encroaching at many points on his own life-sphere, and to wrest his leisure from an environment he was apt to feel as hostile. However, the value of time was for him an affair transcending personal matters. Even when he spoke of the State his thoughts were turned to time. When it was a question of the community, other laws, of course, applied. Here a temporal order to which all were subject held sway, and the individual could not do things or leave them undone as he pleased. The communal order restricted individual time. Such will always be the case. But the humanist measured the worth of the nature of a State by the degree of this restriction. He imagined an order in which the tasks of the citizens were apportioned so sensibly that everyone had time for higher pursuits.

In the *Utopia* the gaining of time was an essential consideration in the establishment of the economic scheme.

> Neque enim supervacaneo labore cives invitos exercent Magistratus; quandoquidem ejus Reipublicae institutio hunc unum scopum inprimis respicit, ut quoad per publicas necessitates licet, quamplurimum temporis ab servitio corporis ad animi libertatem, cultumque civibus universis asseratur. In eo enim sitam vitae felicitatem putant.[55]

The State had to cater for the needs of the individual, and the best State was one which afforded him the greatest degree of leisure for his own intellectual activity. The citizen of Utopia needed to work only six hours a day.[56] In connection with a new ordering of the nature of the State, Campanella too, in his *State of the Sun*, called for an equitable apportioning of leisure time and working hours. The absence of such apportioning seemed to him to be disastrous for the corporate life of the State, illustrating his argument with the example of the city of Naples. There dwelt seventy thousand souls, of whom only ten or fifteen thousand worked. These were broken by the burden of toil; the others, the *ociosi*, met their undoing in the vices born of idleness. The entire structure of the State decayed; every office was inappropriately filled. Against this situation Campanella set his ideal State:

> Ast in Civitate Solis dum cunctis distribuuntur ministeria, et artes, et labores, et opera: vix quatuor in die horas singulis laborare contingit: reliquum temporis consumatur in addiscendo iucunde, disputando, legendo, narrando, scribendo, deambulando, exercendo ingenium et corpus, et cum gaudio.[57]

The assiduity with which the leisure time of the individual was catered for in the ideal State betokened, despite the correlation shown by Campanella to exist between the apportioning of time and the furtherance of matters of state, the ideal of humanity and life rather than the insight of the politician.

There is no place for a demand for leisure in the image of a State conceived by pure political expediency, as for instance in the Platonic State. The humanists no doubt inherited the concept of leisure from antiquity, but they were the first to elevate it from the personal sphere into the political domain and thereby invest it with an importance of another order. The feeling of the citizens of the later Middle Ages for a well planned communal order and the humanistic appraisal of time were doubtless combined in the judgment of the arch-humanist, who was reminded by the belfries of Nuremberg of the value time had there.[58] This value had to be raised to a generally valid principle in the life of the community. Time was one of those possessions in which each must share. In contrast to the abbey of Thélème, which formed an island in the State, but not a State in itself, a self-sufficient communal order must give an important place to the deployment of time.

Given this turn of thought, the desire for time had already

become equivalent to the right to a particular quantity of it, vouchsafed by the State to all. With this the collective demand for time had already been adumbrated in a form it still retains today. But the centre of gravity of this demand had shifted considerably. From the uncommitted literary expression of a day-dream there developed in the course of time an inexorable social exaction in the realm of reality. The humanists spearheaded the intellectual process which culminated in the claim for the eight-hour day. It is a remarkable fact that the question of working hours began to become a social problem at the time of the origins of humanism—that is, at the close of the Middle Ages. In Germany the earliest prescriptions in the matter were formulated in the fifteenth century. In the relations between employer and worker in the Middle Ages time was of such little importance that a master could override it altogether, and might require a journeyman to be ready for work at any hour. Protection of labour did not exist.[59] Thus the demands of the humanists for time do not appear as something detached and separate in the consciousness of that age; they appear not only as a postulate in the purely theoretical sphere, but were also in implicit and unconscious harmony with the concrete and short-term needs of certain sections of the population. The problem of the reasonable and humane division of time within the various communities was no doubt regarded as deserving of the efforts of even the noblest. It was thought that its solution would enhance the happiness of the individual and bring the perfection of the community a step nearer.

It is a commonplace that the historical perspectives of the humanists were clarified and broadened by their preoccupation with antiquity and that an historical consciousness was strengthened which felt itself linked with former centuries in a continuum of human development. It may be that this solidarity with the past, especially one's own national past, expressed itself as a new comprehensive feeling for time in the retention of the traditional spelling of French words also. The advocate of a radical reform of spelling from a phonetic standpoint throws remembrance overboard along with the 'ballast' of letters that have become meaningless. For remembrance resides in a suprapersonal form in traditional orthography. Its disappearance represents a narrowing of the temporal horizon for the sake of considerations of temporary expediency. The loss of the etymological transparency of French words was also one of the arguments advanced by Estienne Pasquier in a letter of 1572 to Petrus Ramus deploring their phonetic spelling:

> Ostez de ceste escriture les lettres que nous ne prononçons pas, vous introduirez un chaos en l'ordre de nostre Grammaire, & ferez perdre la congnoissance de l'origine de la plus grande partie de noz mots.[60]

If we are to designate, in summing up, the imperceptible change in the consciousness of humanity brought about thanks to the merits of the humanists, the essential innovation must not be regarded only as a specific content formulated in a specific way, which simply increased the intellectual riches of humanity, but in addition it must be seen as a new turn of mind and a new feeling for life which expressed themselves in a changed attitude and mode of evaluating things. By virtue of the increasing attention paid to the course and value of time, the medieval naivety and inner unconcern with regard to it disappeared. There now existed a bad conscience where time was concerned. As the Middle Ages moved into the modern era the entire theme of time had its roots in the thought of antiquity. It was through the ancient authors that attention was once again first drawn to time and its importance. Those things which the humanists again brought to bear by virtue of this influence subsequently took their place in the European turn of mind. The new forms of thought seem to have been particularly stimulating to the classical culture of France.

Notes

1 This chapter is—with some modifications—a transcription of an essay of the same title which appeared in the periodical *Die Welt als Geschichte* (Stuttgart, 1941), pp. 165–80.
2 Quoted from H. Rupprich, *Die Frühzeit des Humanismus und der Renaissance in Deutschland* (Leipzig, 1938), p. 76.
3 W. Gent, *op. cit.*, p. 1.
4 *Ep. Mor. ad Lucilium*, I. The subsequent quotations *ibid.*

5 *De brev. vitae*, XII. The subsequent quotations *op. cit.*

6 Albert Counson, 'L'Influence de Sénèque le philosophe', *Le Musée Belge*, VII, p. 132.

7 Here again Seneca is a precursor. Otto Weinreich, *Phöbus, Aurora, Kalender und Uhr* (Stuttgart, 1937), n. 3.

8 Matteo Bandello, *Quaranta novelle scelte* (ed. Picco, Milan, 1911), Nov. 9 (Nov. I, 46 of the edition of Lucca, 1554).

9 H. W. Eppelsheimer, *Petrarca* (Frankfurt a. M., 1934), p. 167.

10 Paul Meissner, 'Empirisches und ideelles Zeiterleben in der englischen Renaissance', *Anglia*, LX, pp. 165f.

11 In a letter written in Latin (Lib. XXIV, ep. I) he gathers a number of quotations from Horace, Virgil, Ovid, Seneca and Cicero, commenting on the brevity of life and the unhaltable flow of time, thus showing how his feeling for time is modelled on ancient concepts.

12 J. du Bellay, *Les Antiquitez de Rome*, I, 5 (Droz, p. 5).

13 Letter of Otto Brunfels to Beatus Rhenanus, 5 February 1520. Horawitz–Hartfelder, *Briefwechsel des Beatus Rhenanus* (Leipzig, 1896), p. 207.

14 L. B. Alberti, *I primi tre Libri della Famiglia*, III (ed. Pellegrini–Spongano, Florence, 1946), p. 255.

15 *Ibid.*, p. 255.

16 *Ibid.*, p. 267.

17 *Ibid.*, p. 268.

18 *Ibid.*, p. 188.

19 *Rime e lettere di Michelagnolo Buonarroti* (Florence, 1908).

20 *Gargantua*, I, ch. 41.

21 *Ibid.*, I, ch. 23.

22 *Opus Epistolarum Des. Erasmi Roter.* (ed. Allen, Oxford, 1960–1934), I, p. 173.

23 From a dedicatory letter of Petrus de Monte Veneti to Andreas Julianus. Sigfrido Troilo, *Andrea Giuliano, politico e letterato veneziano del Quattrocento* (1932), p. 198.

24 Letter of Erasmus to Beatus Rhenanus, St Omer, 13 April 1515. Horawitz–Hartfelder, *op. cit.*, p. 73.

25 Allen, *op. cit.*, I, p. 569.

26 Letter to Barbirianus, quoted from George Ihm, *Der Humanist Agricola* (Paderborn, 1893), p. 17.

27 Alphonse Roersch, *L'Humanisme belge à l'époque de la Renaissance. Études et portraits*, II (Louvain, 1933), p. 150.

28 *Utopia*, dedicatory letter to Petrus Aegidius, *Thomae Mori Opera Omnia Latina* (Frankfurt and Leipzig, 1689; reprinted Frankfurt a. M., 1963, p. 189).

29 Letter of Simon Stumpf to Beatus Rhenanus, 1520. Horawitz–Hartfelder, p. 262.

30 Abel Lefranc, *La Vie quotidienne au temps de la Renaissance* (Paris, 1938), p. 20.

31 W. Pirckheimer, *Apologia seu Podagrae Laus*. Quoted from Rupprich, *Humanismus und Renaissance in den deutschen Städten und an den deutschen Universitäten* (Leipzig, 1935), p. 125.

32 Tobler–Lommatzsch, *Altfranzös. Wörterbuch*, art. *emploiier:* 'Au roi servir ai mon tens emploié.'—*Nîmes* 429; 'ki les desploiiés reploie..., Chil va son tans bien emploiant.'—*Rencl.* C 108, 8; 'Einsi les cinc anz anplea / Qu'onques de Deu ne li sovint.'— *Perc.* H. 6236.

33 Du Bellay, 'Au lecteur', *Divers Jeux rustiques* (ed. Saulnier, Lille and Geneva, 1947), p. 4.

34 Estienne Pasquier, letter to Turnebus, 1552. *Choix de lettres* (ed. Thickett, Geneva, 1956), p. 81.

35 Henri Estienne, *De la Precellence du langage françois* (Class. Garnier), p. 182.

36 Olivier de Serres, *Le Théâtre d'agriculture* (Paris, 1941), p. 113.

37 Montaigne, *Ess.*, III, 13.

38 Lib. XIII, ep. v (Frac. II, p. 233).

39 Lib. IV, ep. III (Frac. I, p. 210).

40 Allen, I, p. 404.

41 Allen, I, p. 536.

42 *Œuvr.*, ed. Moland–Clouzot, II, p. 394.

43 Letter to Hutten, 1519. *Thomae Mori Opera Omnia*, p. 943(*a*): 'muliebre putat, viroque indignum, ejusmodi ineptiis bonam temporis partem absumere'.

44 Descartes, *Œuvr. choisies* (Class. Garnier), II, p. 218.

45 Ed. Foerster.

46 Ed. Luzi, Garzanti, p. 44.

47 Privilege du Roy for Du Bellay's *Deff. et. Ill.* (Roman. Texte, Berlin, 1920), p. 75. Strictly speaking, these are not Du Bellay's words but those of the royal decree.

48 *Anecdotes, Scènes et portraits extraits des Mémoires du Cardinal de Retz* (Tallandier), p. 200.

49 Class. Garnier, p. 132.

50 H. Schweitzer, *Molière und seine Bühne* (Wiesbaden, 1882), IV, p. 89.

51 Roman. Texte, p. 5.

52 Thickett, p. 98.

53 *Œuvr. poét.* (ed. Streicher, Paris, 1958), II, p. 15.

54 Letter to Elisabeth, princesse palatine, 9 October (*Œuvres choisies*, II, p. 258).

55 Lib. II, 'De artificiis'. *Op. Omnia*, p. 206(*b*).

56 *Ibid.*, p. 205(*b*).

57 F. Thomae Campanellae, *Appendix Politicae Civitas Solis* (Frankfurt a. M., 1623), p. 435.
58 F. von Bezold, *Aus Mittelalter und Humanismus* (Munich and Berlin, 1918), p. 148.
59 Alfons Dopsch, review of Schmieder, *Geschichte des Arbeitrechts im deutschen Mittelalter. Geistige Arbeit* (1940), No. 16.
60 Thickett, p. 101.

6
The classical interpretation of time

The view of time prevalent during the Renaissance was the expression of a newly awakened individualism. The private person has a right to his own time, which had to be distinguished from that which he shared with others, i.e. clock time. The enjoyment of this freedom was to be short-lived. The subjective sense of time of the sixteenth century was restrained. In the classical age, time existed not for the individual, but for all; even that most autocratic and absolute prince, Louis XIV, was bound to this universal time. A person was not an individual being but a member of a vast organism, to the coherence of which the integrated observance of time contributed in no small measure. The king precisely determined the course of his day; each of his daily activities was assigned to a set time, and the life of his entourage brought into line with his temporal scheme. This majestic stylization of the royal day was the realization of earlier attempts to make the daily routine of the monarch purposeful and useful as well as exemplary and dignified;[1] this was rendered possible only by the advent of the classical age in France and that of a ruler so strongly conscious of the dignity of his office as Louis XIV. Saint-Simon said of the king that 'Avec un almanach et une montre, on pouvoit dire, à trois cents lieues de là, ce qu'il faisoit'. This rigorous conformity to periods and points of time which one determined oneself had an ethical and aesthetic motivation.

Attempts were now made to curb unrestrained expressions of feeling and to establish a stricter discipline in life. All action must be preceded by reflection, preparation and planning. He who was master of himself was almost master of his time. In order to gain this mastery, all activities, which otherwise would be begun and pursued in an arbitrary manner or according to one's mood or inclination, must be allotted to a time set apart for them. The Jesuit movement advanced along these lines. Ignatius de Loyola prescribed a certain time and duration, established once and for all, for every action and even emotion. Montaigne observed that

> Le repentir qu'il se vante luy en venir à un certain instant prescript m'est un peu dur à imaginer et former.   [III, 2]

The remarkably strong will of this age endeavoured to gain control of the future also; this will had already foreseen and ordained what was to take place in it. This will was not bound to situations; it was not of the nature of a drive, nor did it attempt to bring about the fulfilment of individual wishes in the future, but shaped the future according to a plan and to a considerable extent appropriated it in thought. 'The more an individual's attitude is determined by maxims, the more links are forged between his will, and his whole being, and the future.'[2]

In the tragedies of Corneille we find persons whose heroism seems to place them outside real time. A sense of the flux of time, of regret for transitoriness, of the music of time, of the perpetual recurrence of night and day, and of the irrationality of the future —all this was repressed by these creatures of iron will. Because the hero desires to remain the same person as he has always been, past and future are not differentiated for him by the emotional colouring of a feeling for time.[3] This feeling lies beneath the plane on which he moves. It appeared to the Cartesian mind as material and perceptible and hence of secondary importance. Only those elements of time were perceived which extended into the sphere of intellectual and spiritual values, but not its sensuous manifestations or the manifold forms taken by its moments. One cannot imagine any observation on the subjective length of time, such as 'How slowly the time seems to pass!', coming from the lips of the Cid, of Horace or of Cinna. Time was not an experience which one underwent feelingly and passively, but was a substance shaped by the will.

Descartes gave no place to it in his philosophy. Truth and the recognition of truth lay outside change brought about by time. His thought tended towards the elimination of time from reality. It may even be said that he disliked time. Things which developed over a prolonged period were imperfect. Those things which required time to come into being forfeited their unalloyed beauty and truth. Time took on overtones of contingency, caprice and irrelevance. The philosophy of Descartes stood as a work conceived without precondition or precedent against the creations of time, that is to say, of tradition. It did not build on existing foundations, but started 'from scratch'. A like gesture towards the past was the building of Versailles. Those things which time had made as they were might have taken a different form. In this way the will of the century overcame time in a twofold sense. It subordinated the present and the future to its need to inform time, and minimized those past forces which exerted a constant influence on contemporary life.

The objectivization of time also reflected an aesthetic desire to impart a monumental quality to life. The course of one's external, and to some extent also of one's spiritual, life was seen as something majestic if it was regulated by the calendar and the clock. It resembled the movement of the sun:

Ihre vorgeschriebne Reise vollendet sie mit Donnergang.

The king was a constellation which rose and set. He was a power; and the fact that he was able to make a certain pronouncement concerning time which had not yet become a reality bore further testimony to his might. The precise control of the future is an impressive attribute of authority. In a speech to the Neapolitans, Mussolini promised them that he would again address them in the same place on 24 May 1935.[4] Such promises and commitments raise time to a power standing over humanity, one which makes its irruption into the sphere of personal and individual will.

Time belonged to no one. The king himself had no time, because he was only a sovereign. Time had, as it were, usurped the throne. The king had renounced his personal will once and for all and subordinated himself to the temporal order, which he observed in essentially the same manner as his subjects did. The courtiers' time belonged to the king, even as his own time belonged to his entourage. Saint-Simon formulated this relationship as one of property. With a glance at certain members of the king's entourage, he said of certain times in the course of the royal days and weeks that 'this was their time'.

> C'étoit encore le temps des bâtards et des valets intérieurs, quelque-
> fois des bâtiments, qui attendoient dans les cabinets de derrière.[5]

Or:

> Ces intervalles-là, qui arrivoient trois fois par jour, étoient leur
> temps.[6]

The 'intervals' were pauses in the official life of the king.

The experiencing of time and the surrender to its rhythm and demands appeared to the classical turn of mind as something insufficiently monumental. The subjective attitude of allowing time to flow away in an unregulated manner was eschewed and replaced by a well proportioned temporal scheme. A life, however, that is regulated by the clock may easily sink into banality if it lacks content. The author of the *Caractères* said of Narcisse that

> Il fera demain ce qu'il fait aujourd'hui et ce qu'il fit hier; et il
> meurt ainsi après avoir vécu.[7]

Both court life and the most commonplace activities were regulated by the same divisions of time. The king's watch and the palace clocks showed the same hours as those which governed the life of the ordinary citizen. The fact that one's life was regulated by the clock was, therefore, unimportant in itself, the true touchstone was the significance with which one invested the divisions of one's time. While the strict observance of these divisions might appear to one man as an heroic stylization of life, it might seem to another as a framework destined to remain empty, as a husk without a kernel, as a form with no meaning, justified only by convention. And yet both points of view had a spiritual element in common: in weighty matters, as in trivial, the new view of time aimed at imparting a quality of human generality to human activities and to the days and life of the individual by making all days and all emotions and actions of the same general nature fall in with its norm, thereby simplifying the infinite variety of external and personal circumstances.

To the classical turn of mind, objective time was more than a scheme for ordering matters. It seemed to it to be meaningful. Time had its place in life and art, whereas Rabelais had banished it from both these domains. The unity of time in the theatre can be understood only in the light of the meaning which time as a quantity was now acknowledged as having. As in the later Middle Ages, it was more than a prerequisite of civilization and human intercourse—that is to say, an essentially practical institution—as Bergson considered it to be. It was now an aesthetic formula. The classical theatre insisted that the action of a tragedy be contained within twenty-four hours, and put this principle into effect by reducing, as far as possible, the content of a play to a minimum. The insistence on the unity of time was an abstract formulation. It was not the order and temporal extension of the content of the play which was its chief justification; rather, the content had to adapt itself to the unity of time. This is a standpoint which we noted elsewhere as having a true temporal relevance. Here time took precedence over content. Those attempts to justify the unity of time which postulate the twenty-four-hour period as a prerequisite of dramatic fiction do a disservice to the classical theatre, because they misrepresent its true nature. For the classical theatre, verisimilitude was not a matter of masking unlikely (from the standpoint of time) events with a temporal formula. If this temporal condition was strictly adhered to in the seventeenth and eighteenth centuries, and if its adherents are unable to find a valid and generally meaningful justification for it, this is only because the unity of time was in reality a straining after stylization, irrational and difficult to define. Attempts have been made to prove that the unity of time was an indispensable mainstay of dramatic fiction, whereas in fact it was necessary for stylistic purposes.

The harmonious equilibrium attained by the seventeenth century between the subjective feeling for time and the objective temporal order is comparable to the balance of colours and contours in painting. The individual, also, strove to strike this balance in his own life. Now, more than ever before, time was regarded as a material, in handling which the creative impulse might objectify itself, thereby gaining self-awareness. In this self-created temporal scheme the individual felt himself liberated from nature and from its rhythm. He prescribed his own laws and set his own human time against natural time. It may be that this attitude towards time softened somewhat, in the mind of that age, the harshness which almost always accompanies the experience of transitoriness. Time now appeared less as an inward experience of duration than as something shaped by man. Time was apprehended essentially as a part of one's own self, as something one made of it. The humanity of this century had lost its impressionable receptivity to time, and now applied its will to the mastery and ennoblement of life. Its attitude towards time was now one of activity rather than of passive surrender. The Romantic notion of time imparting a ripeness or scent to things was foreign to this view of time. The awareness of the distance separating humanity from former ages was present in memory, but there was no awe of things past. Time was a valuable material, but one which was not, as it had been in the fifteenth century, prized for its own sake, as something determining the conditions of the struggle of life and its opportunities; nor was it, as in the age of the Renaissance, prized as an experience and a source of joy. New phrases used to denote the value of time were 'the price of time', 'precious time' and 'costly time':

Mme de La Fayette:
Ce prince étoit aussi tellement hors de lui-même, qu'il demeuroit immobile à regarder Madame de Clèves, sans songer que les moments lui étoient précieux.   [*La Princesse de Clèves*, IV]

Corneille:
Le temps est un trésor plus grand qu'on ne peut croire.
[*Rodogune*, II, 2]

Racine:
D'un temps si précieux quel compte puis-je rendre?
[*Bérénice*, IV, 4]

Le temps est cher, seigneur, plus que vous ne pensez.
[*Athalie*, V, 2]

La Bruyère:
Qui considéreroit bien le prix du temps, et combien sa perte est irréparable.   [*Caractères*, 'De la Ville']

Montesquieu:
Après quelques précieux instants d'entrevue.
[*Lettres Persanes*, letter 141][8]

The foregoing examples show that time was prized, not for its utility, but for its aesthetic value. In itself it was neither good nor evil, neither beautiful nor ugly, but the classical spirit saw it primarily as an adjunct to its ideal of a noble way of life. It was a material which one did not merely strive to possess, as the later

Middle Ages did, but also to mould in a seemly and dignified manner. To this turn of mind, time offered a means of shaping life according to its ideal, just as the idea of good works may be associated with the possession of money. This view gave rise to a spirit of temporal economy which lay on a higher plane than the 'time is money' mentality. The person who failed to realize the preciousness of time became an object of disapproval. The modern idea that the time of others is something sacred, one of the noblest fruits of modern civilization, had its source in the social inter-course of the seventeenth century. Respect for the time of others had its roots, not in the individualism of the Renaissance, but in the urbanity of the classical age as a heritage from humanism. This amenity of civilization has its counterpart in the modern polite habit of exorcizing the impatience of those whose time one takes up. Phrases such as 'One moment, if you please' first became current during the seventeenth century.

Molière:
Attendez un moment. [*L'Avare*, II, 4]
Un moment. [*Le Bourgeois Gentilhomme*, III, 10]
Racine:
Arrêtons un moment. [*Bérénice*, I, 1]
Marivaux:
Un moment! [*La Double Inconstance*, I, 4]
Retardez d'un moment. [*Le Legs*, 3]
Je suis à vous dans l'instant. [*Ibid.*, 6]
Vous l'allez voir dans la minute. [*Ibid.*, 7]

The phrase 'One moment, if you please' had another aspect, which, however, did not acquire a more definite form until the Rococo age. It indicated the decline of the heroic, generous and unswerving view of time in the later classical age. The course of the sun and the mode of life regulated by a majestic will could not be influenced by this phrase; but the mutual giving and taking of time, making demands on others' time and acceding to those of one's fellows, both play an important part in the social intercourse of courteous and considerate persons. The mutual giving and taking of time is the very keystone of sociability. This fact had, of course, been realized before the seventeenth century, but now the mental concept of giving up one's own time and making demands on that of others, born of an enhanced social awareness, manifested itself more distinctly in the language.

Whereas the Renaissance had attempted primarily to establish a relationship between man and time, there now arose an enhanced awareness of the function to be performed by time as an intermediary between those who took part in exemplary social intercourse. The endeavour to become a perfect social being—or at least the theoretical problem of the obligations to be met by one such—led to the practical observance or the literary expression of the concept of temporal *bienséance*, of the *égard au temps*, as La Bruyère called it.[9] In *La Princesse de Clèves* this factor governing the temporal basis of human action was occasionally mentioned by name:

Sitôt que le temps de la bienséance du deuil fut passé. [I]

La bienséance lui donnoit un temps considérable à se déterminer. [IV]

Elle résolut de faire un assez long voyage pour passer tout le temps que la bienséance l'obligeoit à vivre dans la retraite. [IV]

The length of a period of time was no matter of indifference in circles where one's life had to be informed as far as possible by dignity and seemliness. The importance of the length of the period in question had to be constantly borne in mind and called for constant vigilance in the deployment of one's time. When La Rochefoucauld recommended that one should keep one's distance in social dealings his precept had a temporal aspect also. To defile time for oneself and others—by selecting an unsuitable moment for some encounter, for example—was barbaric. The ability to pick out the appropriate time for imparting information to others was an important element of social tact.

The 'opportune moment' had, as we have seen, occupied an important place in the fifteenth-century mind. If this consideration again entered the domain of personal interest, it did so in another emotional climate. It was almost exclusively confined to the sphere of social tact, and the 'opportune moment' became a matter of when, and for how long, one should speak or remain silent in company, rather than of choosing a suitable time for action in life generally.[10] The meaning of the term *le monde* likewise became circumscribed at this time; its sense became restricted to 'social élite'. The two opposite poles of behaviour were now to talk *à propos* or *à contretemps*:

> Ce qui est dit à propos est tousiours bon, comme aussi les choses à contretemps ne sont iamais agreables.[11]
> En fin tous leurs discours sont tellement à contre-temps, que les bonnes choses deuiennent mauuaises en leur bouche, et les agreables y perdent toute leur grace.[12]
> Ce qu'ils sçavent, ils ne le iettent pas indifferemment en toutes occasions, et s'ils n'ont lieu de parler fort à propos dans les compagnies, ils aimeront mieux auoir demeuré toute la iournée sans rien dire, que d'avoir dit les plus belles choses à contre-temps.[13]

That which was 'against time' aroused displeasure, for time was more than an external circumstance; it was a finely spun web, which a clumsy person might easily injure. The *homme habile*, however, knew the moment 'that preceded the one in which he might strike a jarring note'.[14] The ability to observe temporal proprieties thus became a yardstick for gauging a person's social worth, and hence also his value as a human being. Social intercourse, as the individual of the seventeenth century understood it, required the observance of a set norm in talking of one's personal circumstances. One's own self must not be allowed undue prominence in point of time. The law-giver in matters of correct social usage drew attention to what was permissible in this respect:

> On déplaît sûrement quand on parle trop longtemps et trop souvent d'une même chose.
> [La Rochefoucauld, *Réfl.*, V, 'De la Conversation']

Prolix discourse—or, indeed, anything of an unduly protracted nature—came to appear as reprehensible, uncivilized or ludicrously foolish:

> Deux nobles campagnards, grands lecteurs de romans,
> Qui m'ont dit tout *Cyrus* dans leurs longs compliments.
> [Boileau, *Satire* III, vv. 43–4]

The society of the classical age may be seen as a model of human community, in which a pronounced individualism was offset by an equally high degree of social responsibility and *bienséance*. 'Individualism' is here to be understood as meaning, not the practical manifestation of a desire for personal power and 'elbow room', but a clear awareness of the scope and limits of one's personal life-sphere. In this society there were no nuclei of personal power with indefinite boundaries; its members were able to raise the distinction between 'mine' and 'thine' from the judicial level to the plane of personal values, to which time belonged. In giving of their time, they knew that they were giving of themselves in doing so. Time was their property, which they sacrificed only to a high ideal of social perfection. The question of the appropriate division of time had acquired a new aspect in the seventeenth century. Whereas in former ages the chief consideration had been one of balancing the interplay of one's activities within the context of one's own life, the equilibrium of temporal relations was now seen as an interpersonal matter. To such thoughts as 'How much time shall I need for this activity, as compared to another?' or 'What is the right way of apportioning time between work and leisure?' was added 'How much time will you allow me, how much of your own will you sacrifice, how much of mine shall I give up?'

This attitude of mind was in harmony with the individualism we have just noted. Time was a currency which circulated among men; it was a medium of exchange. External necessities, however, also exerted an influence on the deployment of one's own time and one's demands on that of others. In this century, persons took leave of one another not only for the sake of good manners but also because their day was filled with diverse business and was planned in advance. They excused their haste on the grounds that time pressed:

Corneille:
Il allait au conseil, dont l'heure qui pressait
A tranché le discours qu'à peine il commençait.

[*Le Cid*, I, 1]

Mme de Sévigné:
Le temps presse.   [6 November 1676]

J'étrangle tout, car le temps presse.   [25 November 1676]

Molière:
Songez que le temps presse.   [*Les Fourberies de Scapin*, II, 7]

Racine:
L'heure me presse: Adieu.   [*Athalie*, I, 2]

La Bruyère:
L'heure presse: il achève de leur parler des abois et de la curée, et il court s'asseoir avec les autres pour juger.

[*Car.*, 'De la Ville']

Marivaux:
Le temps me presse, je suis forcé de vous quitter.

[*La Double Inconstance*, III, 7]

The essential thought underlying such turns of phrase was not the circumstance that they had, in fact, no time, but that they adduced an objective necessity as the cause of their haste, a necessity which they thought to be more cogent than a purely subjective excuse, which might have been expressed thus: 'We have talked enough, I wish to leave now.'

The language of the century shows how time had become something abstract. In poetic usage, terms connected with time lost much of their pictorial quality and their vivid immediacy. With Corneille the word *jour* no longer denoted an experience but became an intellectual concept, just as the words *bras*, *flamme* and *feu* were raised by him onto the spiritual plane. The day was a thing of value which had, so to speak, to be earned, and of which one might be worthy or unworthy. It was thus no longer something morally neutral.

Mais qui peut vivre infâme est indigne du jour.   [*Le Cid*, I, 6]

The day was not only something ethically significant but also something fateful. Days were weighed and evaluated according to their significance for human destinies. They possessed varying degrees of historical import. They possessed differing characteristics—radiance, sombreness, momentousness—which they acquired in consequence of the part they played in the course of events rather than by virtue of their outward aspect or of the impression they created. Let us note some adjectives which Corneille applied to the word *jour*: *funeste* (*Horace*, I, 1), *triste* (*Hor.*, IV, 7), *pompeux* (*Rodogune*, I, 1), *heureux* (*Rod.*, I, 1), *fatal* (*Rod.*, I, 3), *plein de gloire* (*Rod.*, I, 5), *illustre* (*Rod.*, II, 2), *propice* (*Hor.*, I, 2), *criminel* (*Hor.*, V, 2). The natural and the naively self-evident were alien to this turn of mind. Nature was something gross, which must be refined by man.

Accordingly the day was divested of its natural quality and one's own conception of it substituted. Furetière recounted that one asked 'Is it daylight?' when one wished to know whether a person of high birth had risen from bed.[15] The human desire to determine when the day might begin stood against the will of God, who had created the natural day. This human self-sufficiency also accounted for the sudden and arbitrary modification of the meaning of the word *jour* and its conceptual transplanting into the sense of 'facility, means of bringing something about'.[15]

Succession had no real significance for the mind of this century. Temporal relations and periods of time, and duration *per se*, were meaningless in themselves unless they were informed by values. The mere fact of a succession of circumstances was meaningful only if it was relevant to values, and only if a relationship existed between the earlier and the later other than a purely temporal one. Temporal conjunctions were now, more than at any other time in the history of the French language, promoted from their purely temporal significance to the sphere of abstraction and logic. *Pendant que, tandis que, comme, ainsi que, dans le temps que, durant que, cependant que, quand* and *lorsque* were frequently employed in a sense other than the temporal.[16]

The foregoing calls for some comment. Two temporal divisions of a succession and two events occurring one after another can be considered by a person as standing in various mutual relationships. They may be related in time. In that case the consciousness distinguishes what precedes from what follows only as a succession. This temporal interpretation is the basis of logical association. Every relationship of cause and effect contains a temporal element. A difference of value, however, may exist between what precedes and what follows—a third possible relationship. The fact that something or somebody comes first symbolizes the greater worth of that thing or person. This fact, as is logical association, is able to lend a timeless significance to temporal succession. Here it is not time that links events, but the interpretation placed on them. If a difference of time implies a difference of value, then its converse, the simultaneity of two events, must also symbolize equality of value. Let us consider the following examples, both containing expressions of simultaneity:

Molière:
. Habitez, par l'essor d'un grand et beau génie,
Les hautes régions de la philosophie,
Tandis que mon esprit, se tenant ici-bas,
Goûtera de l'hymen les terrestres appas.

[*Les Femmes Savantes*, I, 1]

Racine:
Quand je suis tout de feu, d'où vous vient cette glace?

[*Phèdre*, V, 1]

In the first instance we note that a circumstance is placed on an equal footing with another by being represented as simultaneous with it, and in the second an expression of displeasure at the fact that two states of a different nature and value are obtaining at the same time. It was desired that time should be meaningful; there was, therefore, a tendency to amend or negate mere succession or coincidence if it was void of significance, or even to deny its existence. Events linked by *pendant que* and like words were not neutral as to value. One sought to discern more than mere simultaneity in their temporal coincidence, for otherwise simultaneous events and circumstances would not have been so frequently, and so sharply, set against one another.

Other characteristics of the language in the seventeenth century show how relationships between events were mentally linked and evaluated instead of being merely observed to be connected in time. The preposition *après* and the adverb *enfin* were frequently used in a sense other than the temporal. A succession of events was no longer seen as a process determined by fate, as in the *Chanson de Roland*, nor as a series of spatially juxtaposed points, as in the later Middle Ages, nor yet, as in the time of the Renaissance, was it musically and rhythmically lived through as an experience; it was viewed and evaluated as a meaningful and necessary chain of circumstances. The words *après* and *enfin* often indicated logical consequence and not time:

Racine:
Enfin, tous mes conseils ne sont plus de saison:
Sers ma fureur, oenone, et non point ma raison.

[*Phèdre*, III, 1]

This *enfin*, which Phèdre used in her dialogue with Oenone after Phèdre had vainly attempted to convince Oenone that her

(Phèdre's) desire of setting up a union between Hippolyte and herself must prevail, puts the psychological situation in a nutshell. Its meaning is: 'Do you not see that no arguments can prevail against my love? Do you not see that I need this love in order to live? There is no sense in wasting words.' The word *enfin* draws, from what has been said before, that conclusion which Oenone is reluctant to accept.

Enfin was frequently used in the dramatic literature of the seventeenth century. Before the characters of a tragedy or comedy broke off their speeches or allowed themselves to be interrupted, they often summarized what they had said in a few words prefaced by *enfin*, thus making their position clear. The word often betrayed an egotistic attitude and a certain impatient disregard for another's opinion and point of view. The characters of the classical theatre often pursued a course of action and 'situated' themselves at the same time, by observing themselves from a distance, from a viewpoint other than that of their involvement in action. In this way their speeches did not evaporate from their consciousness as soon as they were uttered; rather, they reviewed what they had said, introducing their summary of it with *enfin*. This word could, of course, refer to what had been said by others or to events. Its use indicated an objectivizing review of what had gone before. What was said or heard did not fly away immediately it left the speaker's lips, but was summarized and a conclusion drawn from it. *Enfin* was of the nature of a realization. When one sees how often it is met with in the works of Corneille, Racine and Molière in particular, one is entitled to regard it as an idiosyncrasy of the classical style. It would be pointless to quote all the examples we have discovered of the use of the word; a perusal of *Le Tartuffe* or *Le Misanthrope* will convince the reader of its frequency. The term indicated that 'the end' was more than a temporal matter of that which came last in point of time; that it was a result, a logical consequence which the understanding might apprehend without further explanation.

The word *après* occurred with equal frequency and with the same shift of meaning:

Molière:
Mais, madame, après tout, je ne suis pas un ange.

[*Tartuffe*, III, 3]

Savez-vous, après tout, de quoi je suis capable? [*Ibid.*, III, 6]

Après son action, qui n'eut jamais d'égale,
Le commerce entre nous porteroit du scandale. [*Ibid.*, IV, 1]

The *après* points to a new situation: this is the way things now stand. It embraces the present from the point of view of the past. It is desired, not to relate or dwell on it, but to draw a conclusion from it. It is used here dramatically, not epically. The characters set out their viewpoint and situation before the spectator. *Après* is the explanation and justification. The temporal element is absent, for it was a question of consequence, not of succession. In this way purely temporal words lost their neutrality, acquiring significance and perspective. The word *après* indicated the importance of an event by pointing to its consequences or by contrasting an event with them. It might be called pragmatic–historical, for it implied an antithesis to the chronicling style of writing, which simply set out events in succession without connecting them in thought. The use of this word is evidence of a way of looking at things that simplified the complexity and irrationality of succession by postulating the premise that the future was the logical consequence of the past. Human behaviour was made to appear more predictable and explicable by an *après tout*. It was an unfree world, in which the past accounted for so much. The possibilities of action were circumscribed. Only one line of action was open to the characters of the classical drama after events of a certain type had taken place:

Corneille:
Je sais ce que l'honneur, après un tel outrage,
Demandait à l'ardeur d'un généreux courage. [*Le Cid*, III, 4]

Thus we see how this century transformed the course of natural time and manifold developments into a system of logical succession and meaningful process. In no other age were so many terms divested of their purely temporal meaning: they were good enough to serve as symbols for higher, timeless truths and values.

The serious turn of mind of the century recognized the nullity of time *per se*:

Nous pouvons regarder le temps de deux manières différentes: nous le pouvons considérer premièrement en tant qu'il se mesure en lui-même par heures, par jours, par mois, par années; et dans cette considération je soutiens que le temps n'est rien; parce qu'il n'a forme ni substance, que tout son être n'est que de périr, et partant que tout son être n'est rien.

[Bossuet, *Or. fun. de Yolande de Monterby*]

From this Bossuet derived the human responsibility of making something of it. For him this meant using it as a means to one's spiritual salvation:

Mais élevons plus haut nos esprits; et après avoir regardé le temps dans cette perpétuelle dissipation, considérons-le maintenant en un autre sens, en tant qu'il aboutit à l'éternité;... il a plu à notre grand Dieu, pour consoler les misérables mortels de la perte continuelle qu'ils font de leur être, par le vol irréparable du temps, que ce même temps qui se perd, fût un passage à l'éternité qui demeure.

[*Ibid.*]

Bossuet was able to illustrate his conception of this ideal use of time *sub specie aeternitatis* by means of the life, of exemplary piety, of the woman whose funeral oration he delivered. His words are a sermon on the Christian deployment of time. The gap separating wasted and well used time is here of existential significance. An individual who wishes to lead a meaningfully disciplined life has taken preliminary decisions as regards time. His will sets itself to the realization of the fixed norms of his ideal of life and in doing so decides once and for all his deployment of time. Bossuet summed up this attitude with the formula of the 'ineffectual abstractum', current since the age of the Renaissance, which characterized the steadfast individual:

Ni les affaires, ni les compagnies n'étaient point capables de lui ravir le temps qu'elle destinait aux choses divines.   [*Ibid.*]

The attempts of various forces to gain control of the human spirit, generally represented in spatial form[17] ('idle things had no place in his heart', 'avarice repressed all emotion within him'), appear, to the mentality which sees spiritual preference for a thing as something which takes up time (the sacrifice of time corresponds to the importance one assigns to a thing), as something competing for an individual's time.[18] The turn of phrase used here—'to steal (take away, deprive one of) time'—had its place in the history of the conception of time. It received the connotation, communicated to humanistic thought by the influence of Seneca,[19] that time is a value and a possession which imposes its obligations.

Notes

1 Cf. Abel Lefranc, *op. cit.*, ch. I, 'La Journée royale de François I à Henri IV'.
2 H. Keller, *Psychologie des Zukunftsbewusstseins*, p. 216.
3 Georges Poulet, *Études sur le Temps humain* (Paris, 1950), p. 97. Poulet closes his chapter on Corneille with the verse from *Suréna* (V, 1):

> Mais je veux que le temps en dépende de moi.

4 H. Keller, *op. cit.*, p. 270.
5 Saint-Simon, *La cour de Louis XIV* (ed. Sarolea, Paris, Ed. Nelson), p. 491.
6 *Ibid.*, p. 495.
7 La Bruyère, *Caractères*, 'De la Ville'.
8 For other examples of *le temps est cher* see Edmond Huguet, *Petit Glossaire des Classiques français du 17e siècle* (Paris, Hachette), art. *Cher*.
9 *Caractères*, ch. V.
10 Castiglione had earlier drawn attention to the importance of temporal propriety as an element of successful social behaviour: 'Però il governarsi bene in questo, parmi che consista in una certa prudenzia e giudicio di elezione, e conoscere il più e'l meno che nelle cose si accresce e scema per operarle opportunamente o for di stagione.'—*Il Cortegiano*, II, 6.
11 *L'Honeste Homme, ou, l'Art de plaire à la Cour*, par le Sieur Faret (Paris, 1637), p. 93.
12 *Ibid.*, p. 149.
13 *Ibid.*, p. 160.
14 La Bruyère, *Caractères*, ch. V.
15 E. Huguet, *op. cit.*, p. 216.
16 Emil Hartmann, *Die temporalen Konjunktionen im Französischen* (Göttingen, 1903), pp. 38, 41, 44, 46.
17 For instance: 'C'est dommage de semer en la terre de notre cœur des affections si vaines et sottes: cela occupe le lieu des bonnes impressions... '—F. de Sales, *Introduction à la Vie dévote*, I, 23.
18 F. de Sales observed, in speaking of pastimes, that: 'si l'on y emploie trop de temps, ce n'est plus récréation, c'est occupation.' —*Op. cit.*, III, 31.
19 Inroads into one's time: '. . . quaedam tempora eripiuntur nobis, quaedam subducuntur, quaedam effluunt.'—Seneca, *Ep. ad*

Lucilium, I. The humanist Dolet reported that he had worked for years on a work entitled *L'Orateur françoys*, 'desrobbant quelques heures... de son estude principalle'.—Marc Chassaigne, *Étienne Dolet*, Paris, 1930, p. 227, n. 1. 'Poterit fortasse Majestas tua me furti incusare, quod tantum temporis quantum ad haec sufficiat negotiis tuis suffuratus sim.'—Francis Bacon, dedicatory letter of the *Instauratio Magna* to James I (*Works*, ed. Spedding, Ellis and Heath, reprinted Stuttgart, 1963, I, p. 123). 'Si ocupada/Vuseñoría está, no será justo/hurtarle el tiempo.'—Lope de Vega, *El perro del hortelano*, I, 16.

7
The shaping of time as a way to happiness in the eighteenth century

The relationship between God and time was, for the Middle Ages, an element of the Christian belief in the omnipotence of the Creator. God stood outside and above time. He might be defined as the Timeless One, for whom years and centuries, past and future, meant nothing. Time was a synonym for humanity, *saeculum*. This constituted a very close link between God and time. With the progress of thought the bond of association between the two concepts became loosened, in as much as time acquired the nature of an independent value in human consciousness. This process took the form of a general secularization of the idea of time, rather than a course of development traceable in the history of philosophical systems. Time itself became a deity whose power was autonomous and independent. The view of time prevailing during the era of the Enlightenment was diametrically opposed to that of the Middle Ages, for in the eighteenth century the powerlessness of God Himself as regards the course of time was affirmed:

> Dieu même n'est pas le maître
> De réformer le passé,
> Le tems promt à disparaître
> L'a dans son vol effacé.[1]

Time had become a law completely independent of God, a law whose workings were subject to no outside control and were therefore inexorable. The powerlessness of God in this respect was shared by humanity. It is true that time was represented allegorically and apostrophized as a person even now, but this personification had become a *façon de parler*. Never did it seem more impersonal and colourless. The eighteenth century was oppressed by the inexorableness of time. The ear of the people of that age did not listen, as the Romantics were to do, to the sad melody of transitoriness, but the human soul was disquieted by the awareness of the impermanence of things. Time was regarded in the main as standing for the limitation, the annihilation and the smallness of human existence. One now became conscious for the first time of the infinity of duration, with all its sobering absoluteness, in which historical existence shrank to an instant. And what of the life of the individual in the context of this brief history of the human race?

Time was something imposing; it was discussed and its power recognized. Well might the new lore of the history of the earth and the world, with its adumbration of vast expanses of time, inspire awe in the minds of men! The characteristics and effects of time passed unnoticed; rather, it was perceived as a horizonless natural phenomenon. Whereas the Romantics apprehended time as the melody and rhythm of becoming, as a personal experience, and as a form taken by the destiny of the individual, during the eighteenth century it was seen as something more or less external. It was an image rather than music, a thing rather than a process. Delille attempted to capture and symbolize the infinity of the ages in various figures of speech:

> L'océan des âges.   [*L'Homme des Champs*, p. 102[2]]
>
> Dans ces fonds qu'a creusé la longue main des âges.
>   [*Ibid.*, p. 102]
>
> Des âges sans fin pèsent sur la pensée.   [*Ibid.*, p. 103]
>
> Combien de temps sur lui l'océan a coulé!
> Que de temps dans leur sein les vagues l'ont roulé. [*Ibid.*, p. 103]
>
> Et le monde, vieilli par la mer qui voyage,
> Dans l'abyme des temps s'en va cacher son âge.   [*Ibid.*, p. 106]

Rousseau asked himself what was really left of life after one had not only seen it in its insignificance in relation to the vast flux of the world, but also after one had made allowances for that time which went to make a human life-time in the arithmetical sense, but of which one was unable to avail oneself in a free and purposeful manner:

> Que nous passons rapidement sur cette terre! Le premier quart de la vie s'est écoulé avant qu'on n'en connaisse l'usage; le dernier quart s'écoule encore après qu'on a cessé d'en jouir. D'abord nous

ne savons point vivre; bientôt nous ne le pouvons plus; et dans l'intervalle qui sépare ces deux extrémités inutiles, les trois quarts du temps qui nous reste sont consumés par le sommeil, par le travail, par la douleur, par la contrainte, par les peines de toute espèce.[3]

This mathematical formulation betrayed a hunger for time, of an acuteness unknown to former centuries. That time which could not be used for the enjoyment of life became an object of dread. The fact that the time granted to man was in fact nothing exercised human consciousness and was a source of inner disquiet to a more marked degree than ever before. The alarm occasioned by the thought of one's lifetime dwindling to emptiness and nothingness accounted for the urgent desire to find that means of deploying time which was alone appropriate. The lament of the age at the brevity of life and the impermanence of things revealed a spirit resembling that of the declining Middle Ages. This spirit, however, was less manifest in the eighteenth century than it had been in the earlier age. The superficial observer may be deceived by a semblance of gladness; yet it was realized how every minute and every day, like the washing of waves, ceaselessly eroded and diminished the substance of human life:

Ainsi l'âge nous presse, et chassant les désirs,
Resserre chaque jour le cercle des plaisirs.[4]

There was no thought of breaking the power of time. The will which defied time wasted its energy. It was useless to pit oneself against time, for everything was subject to its laws.

However, awareness of one's impotence in the face of time did not deter one from turning the time at one's disposal to account as far as possible. Rousseau, who made his distressing calculation of the amount of time actually vouchsafed to the individual, nevertheless found in himself a positive force which enabled him to dismiss the melancholy implications of his reckoning. Away with fearful concern for the short span of time! The individual was too apprehensive of wasting time to be able really to live in it. He had lost touch with the rhythm of the natural way of life. Rousseau now admonished humanity that there was no better method of gaining control of time than to let it flow away. The art of living did not reside in the most comprehensive deployment of time possible, but in the ability to waste it.

> Vous connaissez, dites-vous,le prix du temps et n'en voulez point perdre. Vous ne voyez pas que c'est bien plus le perdre d'en mal user que de n'en rien faire.   [*Émile*, II]

The social individual had lost his sense of the large, natural divisions of time whereby his life was supposed to be regulated and unified. He neglected them, impelled by a desire for enjoyment and by temporal avarice. He was 'toujours pressé de jouir'.[5] In this way he made nothing at all of his time. Rousseau said of such people that

> Ils prodiguent le temps qu'ils pensent économiser, et se ruinent comme les avares, pour ne savoir rien perdre à propos.
>   [*Nouv. Hél.*, V, letter 2]

This was precisely the same view as had been advanced by Rabelais two centuries earlier: the most flagrant waste of time was to count the hours. Thus the authors of *Gargantua* and of *Émile*, so unlike in other respects, were at one in their attitude to time. Both of them took a stand against the prevailing traditional view of time, a view which they saw to have a quality of civilizing inflexibility. With Rabelais the battle cry was 'Freedom for the humanity of the Renaissance!' With Rousseau the fight was led in the name of nature. This fight was more earnest than the smiling Utopia of the Renaissance, for the evil it was desired to combat was greater—if, indeed, to misinterpret and overhear the rhythm of life was in itself an evil.

Humanity, in fact, was in an ambiguous position with regard to time, owing to the circumstance that whilst one avidly sought to have time, one was unable to fill it when one believed one did possess it. Rousseau was the first to warn of the dangers presented by the new European craze for haste and bustle, which cast its shadow over the eighteenth century and set the tone of the current age. Having regard to the fact that his faith in nature was more than a point of view, and that the various aspects of his faith had nothing of the logical consistency one associates with a philosophical system, but had rather that directness and totality of strong conviction and feeling, Rousseau's pronouncements on the position of humanity *vis-à-vis* time were the immediate, and not the indirect, expression of his singular spiritual make-up. He gave vent to his need for emancipation from time, a need which had its roots in his whole general nature. It seemed to him impossible to adapt himself to the demands of the moment as the true social being did. The art of speaking and thinking *à propos*, that quick-wittedness based on the acceptance of an objective time, a temporal order existing between and over individuals—to these Rousseau was unable to adjust himself.

> Mon esprit veut marcher à son heure, il ne peut se soumettre à celle d'autrui. [*Confessions, III (1728–31)*]

He had his own rhythm of life, which failed to attune itself to that of others. His realization of this took the form of a general observation and of a personal need: there existed various attitudes towards time, corresponding to different evaluations of life and modes of living.

Not the least reason for his fondness for solitude was the fact that it afforded him the freedom to live according to his personal rhythm; he expressed himself most clearly on this point in a letter of 26 January 1762 to M. de Malesherbes. His instinct, geared to the antithesis between nature and civilization, enabled him to discern, in all the minutiae of life, the difference between his own view of time and that of the social individual. He detected signs of this difference even in seemingly trivial matters:

> Les lettres des solitaires sont longues et rares, celles des gens du monde sont fréquentes et courtes. [*Nouv. Hél.*, V, note to letter 3]

The time of men of the world was measured; they were unable to write long letters; their existence was divided into an infinitely greater number of parts than that of the natural person; their lives moved to a quicker rhythm and thereby took on an essentially different form. As early as the beginning of the century Montesquieu had recognized the characteristics of his age in the rapid changes of fashion. 'A woman who leaves Paris to spend six months in the country is as *démodée* on her return as though she had been away thirty years,' as Rica, a Persian visitor, observed.[6] Rousseau's criticisms of the manner in which society spent and evaluated its time were made purely from the temporal standpoint—only one of the many from which the 'nature–society' problem might be considered. His hostility towards the traditional 'value of time' was not a subject in itself, and did not represent any pronounced aspect of his nature, but followed the general direction of his thought.

The attitude towards time of the philosopher of Geneva, springing from the new feeling for nature, lived on as a longing in the hearts of his contemporaries, albeit in a less distinct, pronounced and revolutionary form. They strove to attain temporal flexibility in the ordering of their daily occasions. One no longer desired to impose one's strength of will on time, but rather to be led by it, now hither, now thither. Sharp divisions and outlines in the activities and habits of life became blurred. One wished for happiness. In no age was such close attention paid to the 'thermometer reading' of one's own state of happiness at a given time as in this one; no other devised and tested so many recipes, corresponding to such a diversity of needs, for a happy existence. 'Arranger sa vie pour être heureux' was the ideal striven for, but one that was seldom attained as a permanent state.[7] The supreme happiness would, it was maintained, reside in a total unawareness of the passage of time, in the absence of a desire to recall the past or to control the future, that is to say, in an eternal present, devoid of duration or succession.[8] It was futile to attempt to live in such a state of felicity on a permanent basis, Rousseau pointed out, speaking from personal experience. This state of felicity was vouchsafed to the individual only as a rare and exceptional boon of favouring fortune. It was preferable to seek 'ordinary' happiness, more accessible to human beings as they were, and against which circumstances normally set up no barrier. Joy was followed by gloom and a sense of inner emptiness. The desire for happiness ran up against rigid temporal realities. Humanity dreamed of a world with a different order of time. The author of the *Nouvelle Héloïse* added his voice to the chorus which expressed the innermost longing of the age:

> Ah! Qu'on seroit heureux si le ciel ôtoit de la vie tous les ennuyeux intervalles qui séparent de pareils instants!
>
> [*Nouv. Hél.*, I, letter 38]

The malady from which this age suffered was boredom. The goal to be reached by social developments, the rigid time structure of daily life, the elimination of the unforeseen, the knowledge that the future would bring nothing essentially new, that tomorrow would be much the same as today, all this induced an ever-present feeling of *taedium vitae*. Before savouring any pleasure, and indeed before undertaking any activity, it sensed the boredom that would supervene. This gave rise to the general desire to escape from boredom by seeking variety or doing something unusual—but then, what was unusual? The proper deployment of time was a business to which one addressed oneself anew, for

Le bonheur de la vie est dans l'emploi du temps.

[Saint-Lambert[9]]

Such pronouncements were more than mere literary commonplaces or echoings of ancient authors. The apportioning of time was literally a vital issue. One was no longer concerned with dignity and majesty, but with one's own happiness in life and with enjoyment. Delille remarked that the desideratum was

cette voluptueuse distribution de temps entre le sommeil, la lecture des anciens et la paresse.[10]

An ideal method of apportioning one's time between various activities must be discovered. During his sojourn at Les Charmettes the young Rousseau tried out the most varied ways of deploying time, thereby conducting experiments with it, the success of which he gauged by the degree of happiness they had induced in him.[11]

It was also believed possible to attain happiness by restricting one's horizon and living space in withdrawing to an island, which provided spatial as well as temporal insulation from the events of the world. This was the invention of the novel and Rousseau's own experience on occasion. It was a flight from time, whose harsh winds were only too ready to wither the bloom of the illusion of happiness.[12] As early as the age of Louis XIV a need had existed for emancipation from time and from obligations in shaping one's life, but this need lacked the element of seeking and striving, and moreover it had failed to formulate itself in a general way. It was a question of discovering a scheme by which time might be divided and a temporal balance set up between work and rest, earnestness and sport, mind and body—a balance

which would satisfy the demands of all these factors and therefore of humanity. Did such a method of dividing time, which
might impart tranquillity to the course of life and content
humanity by avoiding excess and informing time with harmonious proportions, exist? Was there a temporal order without
those jagged breaks which cut so painfully into the tender and
comfort-loving flesh of the men of this century—'une vie où des
émotions douces et des curiosités fines se succèdent sans vide et
sans heurt'?[13] This ideal was embodied in a linguistic symbol,
now used with new conviction, of the balancing of activities:
tantôt... tantôt:

> Quand pourrai-je, tantôt goûtant un doux sommeil,
> Et des bons vieux auteurs amusant mon réveil,
> Tantôt ornant sans art mes rustiques demeures,
> Tantôt laissant couler mes indolentes heures,
> Boire l'heureux oubli des soins tumultueux,
> Ignorer les humains, et vivre ignoré d'eux.
>
> [*L'Homme des Champs*, p. 145]

This 'now the one, now the other' precluded dedication to the
moment. One did not give oneself up to the present entirely. It
was feared that complete absorption in one activity would lead
one to neglect the ideal of temporal balance. One was not
involved in the Now to such an extent that one lost one's sense
of the impermanence of things. 'Nothing endures!' was the
watchword of the age. The extremes to which the cult of
tantôt... tantôt might be pushed is shown by a sentence of the
seventy-fifth epistle of the *Lettres Persanes*: 'Je crois à l'immor-
talité de l'âme par semestre.' This indicated an abandonment of
perspective. It was no longer desired to take a long-term view of
things. The park at Versailles was replaced by the English
garden, the undisciplined features of which constricted the field
of view, thereby setting up a different relationship—foreign to
classical tenets—between the familiar and the unknown and the
present and the future. The same considerations which applied
in the realm of space also held good in the matter of time. The
familiar, i.e. the present, was required to shrink to a small circle,
while the horizon of the future was extended: it was desired to
familiarize oneself with the latter only by degrees, in small doses,
so that one might be surprised by it. Once again, it was Rousseau

—that great walker—who most plainly perceived this ideal when he expressed his preference for 'le plaisir d'aller'[14] as against 'le besoin d'arriver' when travelling. One's destination must remain out of sight until one actually arrived. One must forget it and savour to the full the pleasure of the moment and of one's slow, step-by-step progress. And, since one's unaided efforts did not enable one to exclude the future from one's mind, obstacles in time were set up and temporal perspective distorted. This was, of course, not a novel and positive way of experiencing time, but an external precaution, comparable to the wearing of blinkers, and which betrayed the incapacity of one's own resources to inform time.

The person whose emotional centre of gravity lay in the present had no need of such negative expedients. The *carpe diem* of Horace was again uttered with a new sense of longing. One strove once again for an enhanced enjoyment of the moment, but one was unable to banish the future from thought. Indeed— and this was an essential characteristic of the age—the phrase was subject to reservations and limitations. The ideal was, rather, a complexity of emotion engendered by damping one's inclinations and disinclinations. One's consciousness was filled with the present, the past and the future, and the supreme skill consisted in setting up the most agreeable relationship possible between them. It was desired to moderate enthusiasms, to tone down strong contrasts and to avoid making harsh decisions, all of which represented jagged breaks in time. It was desired to devote oneself to one form of enjoyment in the knowledge that it would soon be replaced by another. Its end was just as pleasurable as its beginning if boredom were to supervene as soon as it was over. The moment was enjoyed with a glance at other delights which would surely follow:

Delille:
Le lendemain promet des plaisirs non moins doux.
[*L'Homme des Champs*, p. 36]

The greatest boon vouchsafed by time, and one which was particularly desired, was the pleasure of spending it. This was not merely a matter of external circumstances but an end one actively sought to attain. The most pleasant way of passing time

was to be unaware of its passage. The vocabulary was enriched
by new expressions for the art of passing time agreeably:

Saint-Lambert:
Triompher de l'ennui des hivers.   ['Les Saisons'15]

Un facile travail, de doux amusemens,
De la longue veillée abrègent les momens.   [*Ibid.*]

Rousseau:
Mille travaux amusants, qui chassent l'ennui.
                                      [*Lettres sur les Spectacles*, III]

Delille:
Mille heureux passe-temps abrègent la soirée.
                                      [*L'Homme des Champs*, p. 33]

Vieux récits dont le charme amusant les hameaux
Abrège la veillée et suspend les fuseaux.
                                      [*L'Imagination*, I, p. 217]

Combien sur lui du temps pesaient les lentes heures!
Le travail l'abrégeait.   [*Ibid.*, II, p. 113]

The Rococo era wished to have it both ways with time. It was
unwilling to come to terms with it and take it as it was; long
periods must be made short and *vice versa*. Hence one beguiled
time by beguiling oneself. Humanity attempted to 'thumb a
lift' through time; and, since it could not tolerate the wholesome
and natural temporal order striven for by Rousseau, it indulged
in all manner of pastimes. But the ideal in everyone's mind was
that unconstrained attitude towards time. 'Letting the days slip
quietly by' was a phrase frequently used:

Saint-Lambert:
Vos jours, toujours semblables,
Coulent dans des plaisirs, simples, inaltérables.   ['Les Saisons'16]

Montesquieu:
Que les jours coulent agréablement avec toi!
                                      [*Lettres Persanes*, letter 141]

Chaulieu:
Mais, hélas! Ces paisibles jours
Coulent avec trop de vitesse.   ['Sur Fontenai'17]

Léonard:
   Au bord de mes ruisseaux,
J'ai vu couler mes jours comme coulent les eaux.
                                      ['L'Heureux Vieillard'18]

Pierre Blanchard:
Coulez, mes jours heureux, sous ce paisible ombrage.

['Le Rossignol et le Hibou'19]

Rousseau:
La Providence m'offrait précisément ce qu'il me fallait pour couler des jours heureux.

[*Confessions*, IV (1732)]

Ainsi coulèrent mes jours heureux.

[*Ibid.*, VI (1736)]

La douce chose de couler ses jours dans le sein d'une tranquille amitié.

[*Nouv. Hél.*, IV, letter 10]

Quels paisibles et délicieux jours nous eussions coulés ensemble!

[*Rêv. du Prom. solit.*, 10e promenade]

Delille:
Le laboureur en paix coule des jours prospères.

[*L'Homme des Champs*, II]

Tantôt laissant couler mes indolentes heures.

[*Ibid.*, IV]

Time had to be round and soft, so that one might not collide painfully with it. Its function was to lull to sleep the tired humanity of the century and carry it gently from one day to the next. But it was not until the Romantic movement had begun that an attitude of complete and spineless surrender to it was adopted. It would be false to assume that the eighteenth-century view of time was merely instinctive and irrational; it was, however, unreceptive to the nuances and varying experience-content of different periods of time, and to their uniqueness *qua* their non-recurrence.

The human consciousness of that age strove to liberate itself from the cramping circumstances imposed by space and time. One was unwilling to remain where the 'accident of birth' had placed one. The philosophers of the Enlightenment did not see themselves as philosophers of the eighteenth century—that is to say, of a particular period of time—but as the champions of universal reason. Their disinclination to feel at home in one particular age went hand in hand with their taste for that cosmopolitanism which overrode national and linguistic barriers. They were unwilling to acknowledge the fact that every people, every person, every circumstance and every action acquired their peculiar characteristics and individual colourings from the place they occupied in a temporal succession. The Enlightenment's conception of time reduced all those things separated by time to a common denominator of simultaneity, in that it abstracted them from their individual temporal context. In this way the past and the future lost their difference as separate realities. A completely false view of history, arising from a disregard of the time factor, misled the men of the Enlightenment into making prophecies as to the future development of mankind.

Thus we see that two contradictory views of time prevailed during the century. On the one hand, we observe a great deference for time as a natural phenomenon, as a species of physical law; on the other, a complete neglect of or even outright hostility towards time as an historical factor and as a concomitant of development and tradition. Rousseau advocated that the individual should integrate himself with the course of nature; yet he himself, and other exponents of the Enlightenment—if their lives, doctrines and general intellectual attitudes are to be credited—took up a position implying that time imposed no conditions, and was not a binding force and subtly colouring influence. From this disregard of the peculiar qualities imparted by time to all things and events, the ultimate conclusion was drawn by the French revolution: that the temporal element in customs and institutions was valueless and senseless and must therefore be thrown overboard. Differences between modes of thought, institutions and so on were not evaluated according to their place in a scheme of development, but were pronounced to be timelessly just or inequitable, true or false; the past and historical becoming were left altogether out of account.

Notes

1 *Œuvres du Philosophe de Sans-Soucy* (Frankfurt and Leipzig, 1762), I, p. 78 ('Ode au comte de Brühl').
2 Jacques Delille, *L'Homme des Champs, ou, les Géorgiques françoises* (new edition, Paris, 1805).
3 *Émile*, IV ('Beginning').
4 J. Delille, *L'imagination* (ed. Esmenard, Paris, 1806), II, p. 83.
5 *Nouvelle Héloïse*, IV.
6 *Lettres Persanes*, letter 100.
7 Cf. Robert Mauzi, *L'Idée du Bonheur dans la littérature et la pensée françaises au XVIIIe siècle* (Paris, 1960), pp. 36f.

8 *Rêveries d'un Promeneur solitaire*, 5e promenade (ed. Rasmussen), p. 100.

9 From an anthology, *Nouveaux Ornemens de la Mémoire* (Paris, 1811), p. 49.

10 *L'Homme des Champs*, note to the verse 'Ignorer les humains et vivre ignoré d'eux', p. 221.

11 *Confessions*, VI (1736).

12 R. Mauzi, *op. cit.*, p. 651.

13 R. Mauzi, p. 392.

14 *Confessions*, II (Journey to Turin).

15 *Nouv. Ornem. de la Mém.*, p. 14.

16 *Op. cit.*, p. 49.

17 *Op. cit.*, p. 166.

18 *Op. cit.*, p. 271.

19 *Op. cit.*, p. 295.

8
The Romantic feeling for time

The form imparted by Rousseau to the apprehension of time
became more clear and general after his death, in that it imparted
its peculiar Romantic colouring to all departments of life. Relatively to other things, time was now much less an object of the
human creative impulse than formerly. It was desired to refrain
as far as possible from imposing one's will on it. The post-
Rousseau generation was pursued by the idea that time had been
improperly treated. The desire for a salutary and natural relaxation of the tension between man and time led to an active role
being assigned to the latter, while humanity adopted a passive
attitude. One now allowed oneself to be borne along by time
and its rhythm. It was like a mother to whom one turned for
support because one had become unsure of oneself. It was left
to the course of life to fill and inform the hours. This disinclination to use will-power and decisiveness in deploying one's time
occasionally took the form of a conscious gesture:

> Je n'aime pas les gens qui sont si fort les maîtres de leurs pas et de
> leurs idées, qui disent: 'Aujourd'hui je ferai trois visites, j'écrirai
> quatre lettres, je finirai cet ouvrage que j'ai commencé.'[1]

If it be objected that these words were used by Xavier de Maistre
with a comical gesture and are not to be taken seriously, it may
be replied that their value resides in the documentary evidence
they supply of a mental attitude that had not previously existed.
The view was widely held that intellect and will-power had
become atrophied, and that consciousness and judgment were incapable of regulating time as they should, as though in conformity with a natural law to which mind and body were subject
in like degree.

The idea prevailed that a mysterious **force** existed which guided the actions of a man and apportioned his time to better effect than did his intellect, whose infallibility was now suspect. The practice of waiting for the moment of inspiration, of which Musset was a notable advocate, precisely corresponded to this view, which left the choice of the critical moment to forces which were thought to have a greater affinity with time than the normal waking state was credited with. These forces were now allowed to regulate life. Obermann said of sober bourgeois habits that 'Elles remplissent les heures, sans qu'on ait l'inquiétude de les remplir.'[2] He expressed a similar view elsewhere. Once when working in the vineyards he was pushing a barrow to and fro, of which he remarked that 'Il semble qu'elle voiture paisiblement mes heures.'[2] Humanity discarded its preoccupation with the best way of passing time. The question was viewed in a new light. The notions of 'empty time' and 'filled time' formed opposite poles, between which thought now moved. The value of an activity might be measured by its ability to fill time and make an experience of it. In the works of the pre-Romantics we may observe a restless searching for that which might fill time and lend value to duration.

> Je ne vois pas comment je remplirai tant d'heures.
> [*Obermann*, letter 46]

> La prière remplit les heures que le travail me laisse.
> [De Maistre, *Le Lépreux de la Cité d'Aoste*]

The idea that empty time might be filled by something was crystallized by Lamartine into a well defined image:

> Maintenant que mes jours et mes heures limpides
> Résonnent sous la main comme des urnes vides,
> Et que je puis en paix les combler à plaisir
> De contemplations, de chants et de loisir.
> [*Nouvelles Méditations*, V]

No sharp distinction could be drawn between empty and filled time. An approximation to this might be the dividing line between time to come and time already lived through. Any such division, however, was impossible in the case of the latter. Time that had been lived through was *ipso facto* filled time, for it always had content; this content might take the emotional form of a feeling of emptiness and bleakness. The words 'empty' and 'filled' did not refer to time but to the value and significance which a person attached to his thoughts. Indeed, it is likely that the individual was acutely and painfully conscious of the passage of time during such an ostensibly empty period.

One's preoccupation with time was another motive for speaking of its emptiness. For the Romantic mentality, time was a predominant consideration. The Romantic was unable to take a step or utter a word without coming up against it. This was a feature of Romanticism so essential that any account of the movement which fails to consider it must be described as incomplete.[3] The recovery of the capacity for immediate enjoyment and of the awareness of time was an important acquisition. The classical view of time was non-sensual; it had, moreover, exerted a strong influence on the eighteenth century. The difference between the eighteenth century's conception of time and that of the Romantics may be seen in the phenomenon of their divergent view of Italy, whose image and evaluation in the eyes of travellers had always been a sort of intellectual watershed. Charles de Brosses judged Italian works of art, so to speak, in a timeless manner. He wished to see the Colosseum restored; it was thus, in his eyes, to all intents and purposes a latter-day structure. He failed to appreciate that a wide span of time lay between him and the building of the amphitheatre.

But what of the Romantic view of Italy? The sight of the ruins of Rome revive the past in the eyes of a Romantic as nothing else does. The past is spread out before him in a unique sentimental perspective, as though he were surveying it through a lens. A painting of the *Cinquecento* and a contemporary picture give rise to different sensations, if only because they belong to different epochs. Everything is perceived as though coloured by the age to which it belongs. The age of things is not calculated precisely, but vaguely apprehended; yet it is strongly felt. Everywhere the past obtrudes on one's consciousness. Bonaparte spoke in Egypt of the millennia which looked down upon his soldiers. The various ages had their individual shades of colour. The same thought was expressed by Mme de Staël when she spoke of the obelisk in St Peter's Square as 'an awe-inspiring contemporary of so many centuries'.[4] A retrospective glance over the centuries induces a feeling of giddiness. One may be on familiar terms

with a far-distant age; one's intellect may set it in its proper
historical perspective and estimate its remoteness from the present
day without any emotional involvement in the vast ages separating it from ourselves. The eighteenth century saw things in this
unencumbered temporal perspective. It leapt over what lay
between itself and earlier ages, as one skips the chapters of a
book. The Romantics, however, did not jump over the intervening time in this way but, as it were, made their way through
it. When they transplanted themselves into the spirit of the
Middle Ages they carried the burden of the intervening centuries
as they did so. It was an actual effort for them to make their way
through the past in order to arrive at an earlier age. They swam
against the stream of time. The use of the word *remonter* in this
sense reflected the idea:[5]

Rousseau:
Il faut que je remonte au moins de quelques années au temps où,
perdant tout espoir ici-bas, etc.
[*Rêv. du Prom. solit.*, 2e promenade]

De Vigny:
Quand je remonte à mes plus lointains souvenirs.
[*Servitude et Grandeur Militaires*, I, ch. 3]

Lamartine:
Remontons le cours des années.
['Le Passé' (*Nouvelles Méditations*)]

Hugo, introducing his novel *Notre Dame de Paris* with the
words 'Il y a aujourd'hui trois cent quarante-huit ans six mois
et dix-neuf jours que...', challenged the reader mentally to
traverse this period of time and to place the story in its temporal
perspective.

The past is something one cannot evade, not only because one
sees traces of it at every turn, but also because one carries it
about with one in memory as an inalienable possession. The
individual perceives that his life is twofold: on the one hand there
is one's earthly life between birth and death, and on the other
the passing of the centuries, the sound of bells reaching one from
a lost age, an age that comes alive with the vividness of a personal
memory. In the Romantic period one's own past weighed more
heavily on one's mind than formerly. One moved forward with
one's head turned backward; time, like an ever-growing shadow,

accompanied one. Past experiences became present and were represented as such by the use of the present tense:

> J'entends encore mon père tout irrité des divisions du prince de Soubise et de M. de Clermont; j'entends encore ses grandes indignations contres les intrigues de l'Œil de Bœuf, qui faisaient que les généraux français s'abandonnaient tour à tour sur le champ de bataille, préférant la défaite de l'armée au triomphe d'un rival; je l'entends tout ému de ses antiques amitiés pour M. de Chevert et pour M. d'Assas, avec qui il était au camp la nuit de sa mort.
> [Vigny, *Serv. et Gr. Mil.*, I, ch. 1][6]

This use of the present tense in speaking of past events was of a different nature from the old device of recounting a story in the present tense: 'Le corbeau ne se sent pas de joie.' Here the Romantic was not transposing a story or a past event into the present tense merely by using a stylistic artifice; he did not exclude the time separating him from the event he was recounting from his mind; this time was fully present in his consciousness. The phrase 'I can still hear him' did not obliterate distance in time; he saw the event through the intervening time as clearly as if it were taking place before his eyes. The past had the immediacy of the present, but it was apprehended by that side of the author's spiritual make-up which was turned towards the past.

Time had lost its abstract quality; it had become something visible, an identifiable and powerful force. The course of time, the flux of the centuries, the decay of buildings and the rank growth of vegetation were acoustic and optical impressions, for which the poets had a keener ear and eye than ever before. They developed an auditory perceptiveness, of a hitherto unknown acuteness, for the ceaseless trickling away of time. Along with this, the ethos of time, its divine compensatory function, was recognized. Time revealed the nothingness of all things. With Chateaubriand, the awareness of this took the form of *ennui*, of indifference towards everything that was subject to time.

In Lamartine's famous poem *Le Lac* the feeling for time was paramount. The subjective sense of servitude to time, the awareness that time was the most fundamental experience of man, here emerged triumphantly. This poem bears witness to an experience of time more subjective than can be found in any other

work of French literature. Time had lost its contours; it was no longer something stable and tangible. It had ceased to be extension, and was therefore devoid of quantitative value. A period that might seem like a minute to one might appear as an eternity to another. An hour, a day, a year, were nothing as far as their association with the clock and the calendar was concerned. The calendar, in its strict sense of fixed temporal units, was revalued and thrown into the melting-pot in this poetic recasting of the idea of time. They had all become dynamic; they had acquired a high degree of mobility. Let the reader enumerate the terms connected with time occurring in the poem and then consider the way they are used! 'Dans la nuit éternelle emportés sans retour'; 'l'océan des âges'; 'l'heure fugitive', 'le temps n'a point de rive, il coule'; 'moments d'ivresse'; 's'envolent loin de nous de la même vitesse que les jours de malheur'. One feels forcibly carried along by this current, powerless to resist it:
The poem

> O temps, suspends ton vol! et vous, heures propices,
> Suspendez votre cours!
> Laissez-nous savourer les rapides délices
> Des plus beaux de nos jours!
>
> Assez de malheureux ici-bas vous implorent:
> Coulez, coulez pour eux;
> Prenez avec leurs jours les soins qui les dévorent;
> Oubliez les heureux.
>
> Mais je demande en vain quelques moments encore,
> Le temps m'échappe et fuit;
> Je dis à cette nuit: 'Sois plus lente'; et l'aurore
> Va dissiper la nuit.

abounds in words to do with time: *nuit, éternité, âge, jour, année, soir, temps, heure, rapide, moments, lent, aurore, fugitif, vitesse, éternelle, passé.* They had now become stripped of their *sens propre.* Their meaning is vague, general and blurred. Terms which in normal practical use delimit one another sharply, and which can be measured by, and divided into, one another, because they stand for invariable divisions of time, are here interchanged and merge into each other. There are no longer any boundaries between days, hours and moments. The only thing that matters here is their essential quality, i.e. that they pass away. They are, like other terms in the poem, dissolved in rhythm. It is only by virtue of the fact that the terms used are predominantly auditory elements, and that their conceptual content is vague, colourless and attenuated, that they recur so frequently. The linguistic symbolism of the poem conveys the message that time is present in everything.

The conceptual use of the word 'time' is not uniform. The classical language had always confined itself within the limits of a narrower and more distinctly recognizable field of meaning. It used the word logically and in a manner open to *la raison*. Here its function is largely that of emphasizing the import and depth of experience. The word has the effect of a catalyst with which everything that was said was treated, in order to lend it weight and colour. Time served as a new and hitherto unused symbol for freshly discovered regions of the subconscious, the incomprehensible, the obscure, the other-worldly and the alien.

In the language of the Romantics the spatial world was turned into something temporal. Juxtaposition in space was seen as succession in time. Victor Hugo expressed the fact that seven pillars stood in the courtyard of the Palais de Justice in Paris in these terms:

A quelques pas de nous, un énorme pilier, puis un autre, puis un autre; en tout sept piliers dans la longueur de la salle.
[*Notre Dame de Paris*, I, 1]

What we shall call the 'Romantic plural' was also a subjective extension of the world. In Victor Hugo's line

C'est Dieu qui les suspend en foule aux cieux profonds.
[*Feuilles d'Automne*, XXXV, 1]

cieux is an expression of limitlessness rather than something throwing light on the cosmology of the poet. 'The heavens' do not stand next to one another as a definite number of physical entities with clearly defined outlines, but form a series in the eyes of the poet. The plural extends the fixed, resting world, which can be embraced at a glance, into the domain of time. The idea of depth, which one gains only by the act of penetration (in the mental or material sense), strengthens this subjective 'temporalization':

> A ces mots, le météore rentra dans la profondeur des cieux, l'aérienne divinité se perdit dans les brumes de l'horizon.
> [De Maistre, *Expédition nocturne*, 34]

The Romantic attenuated and blurred the corporeal nature of things by characterizing them in the plural:

> Tout à coup une singulière harmonie résonna dans mes solitudes.
> [Gérard de Nerval, *Aurélia*, I, 8]

These 'solitudes' might be walked through as though they were space. Their boundaries receded ever farther from the one who traversed them.

The essential quality of things is their lastingness; the most important feature of terms is their temporal significance. The Romantic movement recognized and developed those meanings, susceptible of application in the temporal sense, that were inherent in certain words. In the eighteenth century 'Gothic' was a term denoting a stylistic quality associated with medieval works of art. For De Brosses, the terms 'Gothic' and 'Classical' formed a pair of opposites equivalent to the antithesis of 'ugly' and 'beautiful'; that is to say, their historical relationship was overlaid by a differentiation of aesthetic values. In the early nineteenth century, however, the word 'Gothic' acquired a definite historical significance in addition to its aesthetic one, in that it denoted a particular epoch in history.

The awareness that the individual's time was now his own property had now really made itself felt for the first time; it took the form of an insight into the relationship between man and time rather than of a fear of political commitment and involvement. A person's time was his own; it formed part of him, of his ego; he did not live in a general and objective time, but in his own personal and inalienable one. The awareness of this was reflected in the silent and unobtrusive workings of linguistic change. The following phrase from a contemporary author:

> Et sans bien savoir à quoi j'emploierais ma nuit,[7]

indicated, by its use of the possessive pronoun, that time was in some sort the property of the individual. It was not the night as a universal phenomenon that was meant, but rather one's own night, shared with nobody, because everyone lived through his own time.

The use of the possessive pronoun in this special sense did not occur until the language had reached an advanced stage of development. In the Middle Ages *mon jour* and *mes heures* were not as yet used to denote certain specified days or hours. A day or an hour was not seen as a person's property. It was the individualism of the Renaissance, which imposed on the individual the duty of shaping time by his own decision, that first used the possessive pronoun as a clear formulation of the idea of possession. Rabelais said of his hours that 'Je fais des miennes à guises d'estrivieres.'[8] The seventeenth century had no conception of time which no one could share with another, for in that age time was common property. Only with the advent of the eighteenth century was this 'property relationship' emphasized.

Since the individual's ego now seemed to him so interesting and valuable, the feeling that one's own time formed an important part of one's self was strengthened. The possessive pronoun became more frequent.

Delille:
Laissant couler mes indolentes heures.
> [*L'Homme des Champs*, p. 145]

Rousseau:
Mon après-midi se passa dans ces paisibles méditations.
> [*Rêv. du Prom. solit.*, 2e promenade]

Obermann:
Il semble qu'elle voiture paisiblement mes heures.   [Letter 9]

Goethe:
A young heart is inseparable from a maiden, and spends all the hours of its day in her company.   [*Werther*, I (27 May)]

A Romantic, who listened to the rhythm of his effective life in solitude, realized that his day and his hours had little to do with the 'public' day which regulated the course of the outside world.

Musset:
Les songes de tes nuits sont plus purs que le jour.   [*Rolla*, III]

Pour lui payer sa nuit.   [*Ibid.*]

La femme la plus sèche et la plus malhonnête
Au bout de mes huit jours trouvera dans sa tête,
Ou dans quelque recoin oublié de son cœur,
Un amant qui jadis lui faisait plus d'honneur. Etc.

[*Namouna*, I, vv. 38–41]

Pourquoi promenez-vous ces spectres de lumière
Devant le rideau noir de nos nuits sans sommeil.

[*Ibid.*, vv. 58–9]

Il les vit, le jour venu, après douze heures employées à boire et à remuer des cartes, se demander en faisant leur toilette quels seraient les plaisirs de leur journée.   [*Frédéric et Bernerette*, IV]

It was possible in Old French to use a possessive pronoun as a prefix to a word to do with time. One often said *ma vie, mes jours, ma jeunesse, nos ans*, but all these meant 'life' or 'age of life'. *Nos beaux jours* meant nothing new if used in this general sense. The innovation was that a particular period of time was, independently of the idea of a person's life, attributed to him in the sense of a possession which he might put to whatever use he thought fit, and also in the sense of a fateful inner relationship between the two. If we examine the affective content of this new linguistic usage more closely we detect an awareness that there were as many ways of experiencing time as there were persons, and that time was not the same for all, but that each had his own. The individual clothed himself in it as though in a garment especially made for him. Whereas in the classical era there existed a particularly acute sense of inter-personal temporal relations, because the attention of that century was fixed on everything that was socially serviceable, time as a binding force, as a supra-personal ordering factor, *le temps social*, now faded in consciousness. That which bound men to one another was not the injunction to share a community based on temporal togetherness but the feeling of belonging, in a manner determined by fate, to the same generation, to the same moment of history. In the classical age the emphasis was on intercommunication and simultaneity; now, however, the time-lines of individuals resembled threads that were only loosely woven together. The Romantics were, it is true, familiar with simultaneity, though not as a clear external relationship, but as a feeling of emotional solidarity.

The individual felt utterly subjugated to time. He saw himself being carried along with other men and things in its swift current, at once uncontrollable and unpredictable. He was helpless and knew it. Time was the most powerful, the most inexorable and the most enigmatic force to which he was subject. He therefore felt it—and that much more intensely than formerly— as something living and personal. The classical poet saw time as one force amongst others. It occupied a place in the company of honour, love, heroism and the sense of duty. It stood in the background as an almost invisible figure, of whose workings humanity was only occasionally and fleetingly aware. With Corneille, its effects were apprehended intellectually rather than emotionally. Not so with the Romantics! Its inexorableness pursued them always. They saw their fateful subservience to time in everything they experienced and contemplated. This realization played a part in setting the mood and atmosphere of Victor Hugo's poetry, but it was by no means confined to that writer's work. The individual now encountered time in his every experience. Those things and events he met with were considered in relation to time; this relationship often imparted a peculiar quality to them.

Love was a case in point. It influenced the individual's sense of time in such a way that he saw the past and the future in a changed light.

> L'amour n'est qu'un point lumineux, et néanmoins il semble s'emparer du temps. Il y a peu de jours qu'il n'existait pas, bientôt il n'existera plus; mais, tant qu'il existe, il répand sa clarté sur l'époque qui l'a précédé, comme sur celle qui doit le suivre.
>
> [Constant, *Adolphe*, ch. 3]

Love seemed to endow one with the ability to extend the moment both backwards and forwards in time. Longer or shorter periods of time were for the Romantic mentality not a means of ordering thought or life, but represented experiences, pleasurable or painful, which left their imprint on the mind.

That same time which embraced all seemed to appear in a thousand different forms, depending on the circumstances prevailing at the time. It seemed to Obermann, as he stood on a lonely mountain peak, that time passed more slowly there than elsewhere:

Mais là, sur ces monts déserts, où le ciel est immense, où l'air est plus fixe, et les temps moins rapides, et la vie plus permanente.
[Letter 7]

Les heures m'y semblaient à la fois et plus tranquilles et plus fécondes, et comme si le roulement des astres eût été ralenti dans ce calme universel, je trouvais dans la lenteur et l'énergie de ma pensée une succession que rien ne précipitait et qui pourtant devançait son cours habituel. Quand je voulus estimer sa durée, je vis que le soleil ne l'avait pas suivie. [*Ibid.*]

With other peoples and in other countries, too, time took on a different aspect. Mme de Staël observed that in Germany not only did other customs and values, another intellectual and spiritual atmosphere, hold sway, but also a different time, which at first she found oppressive.

Il semble que le temps marche là plus lentement qu'ailleurs, que la végétation ne se presse pas plus dans le sol que les idées dans la tête des hommes. [*De l'Allemagne*, I, 1]

These observations in the sphere of time induced an enthusiasm akin to that engendered by a scientific discovery. One listened to one's own heartbeats and measured them against time, which sometimes seemed to pass too slowly and sometimes too rapidly.

Slowness exerted a peculiar fascination on the Romantic temperament. Leisureliness and slowness were in sympathy with the dreamy, wool-gathering Romantic soul, to which haste and speed were fundamentally antipathetic, because they militated against the cultivation of 'soul states'. Indeed, it may be said that a taste for slowness developed. The word 'slowly' acquired new overtones in the pages of Rousseau:

Je gravissais lentement et à pied des sentiers assez rudes.
[*Nouv. Hél.*, I, p. 23]

Je me laissais aller et dériver lentement au gré de l'eau.
[*Rêv. du Prom. solit.*, 5e promenade]

The Romantics savoured slowness with a sort of voluptuousness and melancholy earnestness:

Mme de Staël:
Il marchait lentement à côté de Corinne; l'un et l'autre se taisaient.
[*Corinne*, IV, ch. 3]

Vigny:
Ce chœur mélancolique s'élevait lentement.

[*Serv. et Gr. Mil.*, II, ch. 4]

Ces brouillards s'épaississent lentement.   [*Ibid.*, II, ch. 4]

Des germes tout nouveaux qui croissent lentement.

[*Ibid.*, III, ch. 6]

Hugo:

l'ombre du tombeau
Se leva lentement sur son visage blême.

[*Les Châtiments*, V, 11]

Gautier:
Deux fois nous montâmes l'escalier lentement, comme si nos bottes
eussent eu des semelles de plomb.

[*Souvenirs Romantiques*, 'Victor Hugo', première rencontre]

A particularly strong impression was made by the sound of bells
dying slowly away in the night:

De Maistre:
L'horloge du clocher de Sainte-Philippe sonna lentement minuit.

[*Expédition nocturne*, ch. 37]

Musset:
L'horloge d'un couvent s'ébranla lentement.   [*Don Paez*, I]

The course of the world had, naturally, not slowed down, but
the more leisurely phenomena provoked contemplative thought;
they had atmosphere and induced a dreamy mood. The epithet
'slow' indicated an essential quality of things. Slowness is not
only a degree of speed in the technical sense, something one can
read from a speedometer, but has also unmistakable spiritual
qualities: gravity, pensiveness, grief and gloom. The Romantics,
who were inclined to weep rather than to rejoice, recognized the
spiritual significance of slowness. This quality readily acquired
overtones of the unfamiliar, the uncanny and the admonitory.
This aspect was noticed by the Symbolists in particular. There
was now a fondness for emphasizing the temporal extension of
processes and states in order to lend greater weight to them and
thereby to accentuate their emotional content. An indication of
the length of an action often imparted enhanced intensity to it.
The Romantics achieved this effect by means of the simple
stylistic device of using an adjective attributively:

Sénancour:
Voici trois jours qu'un pauvre estropié et ulcéré se place au coin d'une rue tout près de moi, et là il demande d'une voix élevée et lamentable durant douze grandes heures.   [*Obermann*, letter 10]

Sainte-Beuve:
Les deux personnes qui venaient occuper cette humble et assujettissante position, et passer de longues journées sans murmure à ces fenêtres monotones.   [*Portraits de Femmes*, 'Christel']

Six longs mois s'étaient écoulés depuis la première visite.   [*Ibid.*]

Hervé, attentif et discret, vint, revint, et s'y trouva naturellement assis, chaque après-midi, pour de longues heures.   [*Ibid.*]

Thus the attitude of the Romantics to time was, like that of the sixteenth century, again a subjective one. However, these attitudes were dissimilar in other respects. The Renaissance turned to time in a youthfully optimistic spirit, often shaping it according to the whim of the moment, and was disinclined to take heed of the morrow. But if it was not 'eager' for the future, it nevertheless viewed it with optimism and confidence. It was so sure of it that it was able to dispense with laying the foundations, in its emotional and general outlook, for that small future represented by 'tomorrow' and 'the day after'. The Renaissance's sense of the future revolved about the word *postérité, les siècles futurs* and the like. This sense of certainty with regard to the future had become objectified, in that it had attempted to hold its own against other streams of thought (*La Querelle des Anciens et des Modernes*), and during the eighteenth century it increasingly took on the nature of a political issue. During the great revolution, long-cherished expectations were to be fulfilled; the potentialities of the future were to become a reality. This occasioned a sobering of the general consciousness. Men were *désillusionnés*—and acutely aware of it. One felt as though one belonged to a posthumous and futureless generation. The Romantics did not look forward, but backward. The phrase 'l'esprit tourmenté du passé et attendant peu de chose de l'avenir'[9] was a striking characterization of this emotional outlook.

The drama of the Romantics was also influenced by their conception of time. The unity of time, which for Voltaire had still been something dictated by common sense, lost its meaning. People were now unable to see any usefulness in this objective, empty and schematic temporal stipulation. The true experience of dramatic action was not a matter of the clock. For one under the sway of love, hope, fear, despair, jubilation or rage, twenty-four hours might seem as a moment or an eternity. Since the Romantic had no sense of time thought of as something external and spatial, disregard for the unity of time did not affect the illusion produced in him by the play. Every person and circumstance had its own time. It was the dramatist's privilege to use the stage as a medium for projecting his conception of time. Indeed, the extent to which he was able to transmit his own temporal rhythm to the spectator was a measure of his poetic abilities. The view now gained ground that poetry had nothing to do with measured time; the clock and the poet were mutual enemies.[10] Every event and circumstance entered the sphere of experienced duration as soon as true poetry had touched upon them. With the French Romantics, the exclusion of clock time from the stage and from verse was part of their general anti-classical attitude and of their unwillingness to be bound to the rules rather than a clear awareness of the new view of time.

Those minds emancipated by a study of Shakespeare made a broad-minded view of the role of time in the drama their own. 'Toute action a sa durée propre comme son lieu particulier. Verser la même dose de temps à tous les événements! Appliquer la même mesure sur tout!' exclaimed Victor Hugo.[11] Manzoni and Alfred de Vigny expressed the same view.[12] This did not represent a general view of time, but one restricted to the exigencies of the theatre. In this respect the Romantic movement differed from the Renaissance, which had found its most eloquent exponent of the new sense of time in Rabelais. The dictum of the author of *Gargantua* on the relationship of humanity to time was formulated by the Romantics as the preserve of the poet: 'D'abord il prendra dans sa large main beaucoup de temps.'[13] A comparison with the Renaissance renders evident the limited scope of these new formulations, for whereas in the earlier age the conception of time had embraced life in general, it now applied only to the theatre. And yet a new relationship to time, established in the general consciousness, underlay this literary *parti pris*, even if the Romantic did not appear to realize its broader implications. The source of his hostility to the classical temporal unity lay in his new view of time and not in a new conception of the theatre *per se*. If the dramatic poet should 'grasp much time in his broad hand', this implied that time must not be measured, and that no quantitative limitations existed between time and humanity. Apart from its subjection to the natural temporal order and to the rhythm controlling all forms of life, humanity was bound by no conditions in its relationship to time. It was free to shape it at will.

The Romantic insistence on local colour was likewise bound up with its new conception of time. This local colour might equally well be called *couleur historique*, for the poetic ideal was not only to capture strictly local colour, but also to achieve an historical accuracy unblemished by anachronisms. In their Renaissance dramas Musset and Hugo were at pains faithfully to bring out the age in which the plays were set. They were aware of the truth that every historical epoch had its individual countenance. They had a sense of the infinite variety of phenomena in time. Everything was subtly modified by the age to which it belonged. The absolute did not exist. The Romantics endeavoured to see men, circumstances and values as the unique products of their respective historical epochs.

This must not be understood as implying that they possessed impeccable historical perceptivities. Their ability to appreciate that different ages had their various nuances of colour did not necessarily mean that they saw every epoch in its true light. Nevertheless, the realization that everything was essentially characterized by the age to which it belonged, and that it received its unique colouring and incommunicable atmosphere from it, was a valuable step forward. It is true that this realization was never reduced to a clear formula *sub specie temporis* by the Romantics. Lamartine, in speaking of the link between spiritual values and their origins, attached as much importance to their place as to their time of birth.[14] But when one spoke of nature, which had produced something sublime, or of the place of an act of intellectual creativity, then these were placed in their temporal context also. Just as things now juxtaposed in the present were seen as stratified in time, so, conversely, succession was seen in the form of an extended and varied landscape. History became a geography of the centuries; mountains, valleys and climates were seen as events in time. When Jean-Jacques Ampère spoke of the *carte historique* of the Orient,[15] he did not have an historical atlas in mind but the hills and valleys of the rise and fall of the successive cultures of the East. This way of looking at things imparted to the past a reality like that of the present. The sixteenth century, for instance, resembled a still existing mountain range with peaks of varying heights.

The relationship between the present and the past became changed. The existence of many hitherto concealed links between the remote past and the present day was sensed and revealed, as a water diviner detects the presence of an underground spring. A new awareness of the factors separating the men of different ages gave rise to a preoccupation, so far unknown, with weaving all manner of connecting threads between former ages and the present. The Romantic consciousness, by dint of absorbing all these different elements of the past, lost touch with the present to a remarkable degree. The line of demarcation between the real and that which was no longer so became blurred. Indeed, this extension of the present and the imparting to it of a poetic quality by reminiscences of things past was an essentially Romantic trait. One's consciousness was open to the past, which was allowed to invade it at every opportunity. Time was no longer coloured simply by its significance for individual destinies, an idea which found its expression in such phrases as *le jour funeste, l'heure fatale, le moment heureux*.

The fact that so many events occurred at once lent complexity to the moment. Whereas the classical mind saw related values in simultaneity, and whereas temporal coincidence often seemed to it to be something external, fortuitous and meaningless, the Romantics apprehended simultaneity as a mood which combined the most varied impressions into a single mental state, that of awareness of the 'Now'. A given moment of time in its passage was, so to speak, apprehended in several dimensions. A circumstance was not only a link in a chain of successive events, not only a point on a line, but went to form, together with other simultaneous events and circumstances lying outside this line, an integral state of consciousness. A new power of synthesizing events that were simultaneous, and therefore apprehended as a whole, enlivened Romantic literature and was put to use in the historical novel as in lyric poetry. Simultaneity was a uniting force, whereas temporal differences were a divisive one.

The preposition *avec* provided a linguistic means of symbolizing the solidarity of events in different spheres by the idea of simultaneity:

> puis enfin sa maladie de langueur qui l'emporterait, disait-il, avec les feuilles d'automne, et qui le coucherait au cimetière de son village. [Lamartine, *Raphaël*, prologue]

With *avec* more than the external fact of simultaneity is here conveyed. It emphasizes—in contrast to a phrase that expressed only simultaneity, e.g. *en même temps que*—a community and togetherness of destiny, an inner association between the man and the falling leaves. In this usage the word *avec* received temporal connotations by the fact that it linked processes, not persons and things:

> Le vent était tombé avec le jour. [*Raphaël*, XXI]
>
> Ces noms avaient monté avec sa colère.
> [*Histoire des Girondins*, I, 19]

In this way the effects of time were over-estimated. A person who had reached a particular age of life had more in common with his coevals than with his own self, considered as belonging to an earlier period in his life.[16] Parity of age was a source of mutual sympathy and complicity. This was reflected in Musset's *Enfant du siècle*. Here we see the origin of the now current obsession with one's own generation. The Romantics imagined and portrayed themselves as children of their age: one is reminded of the verses in which Hugo incorporated his birth year into the course of history:

> Ce siècle avait deux ans, Rome remplaçait Sparte,
> Déjà Napoléon perçait sous Bonaparte.
> [*Les Feuilles d'Automne*, I, vv. 1–2]

The effects of time on external circumstances and things were only a faint reflection of the changes taking place within oneself. No two hours of one's life were alike.[17] One's mental horizon changed more rapidly than the seasons.

It was, however, possible to arrest the flow of time in imagination by seizing a moment from its ceaseless flux and clinging to

it. In the Romantic age this device was often resorted to, a
frequent introductory formula being *C'est l'heure où...* Döhner
considered this phrase to be no more than 'an artifice of method'.[18]
It served to create a point of repose and a moment of reflection:

> C'est l'heure où, sous l'ombre inclinée,
> Le laboureur, dans le vallon,
> Suspend un moment sa journée,
> Et s'assied au bord du sillon;
> C'est l'heure où, près de la fontaine,
> Le voyageur reprend haleine
> Après sa course du matin;
> Et c'est l'heure où l'âme qui pense
> Se retourne, et voit l'espérance
> Qui l'abandonne en son chemin.
>
> [Lamartine, *Le Passé* (Secondes Méditations)]

The phrase served also to emphasize the uniqueness and particular significance of the moment. Time was banished. Romanticism,
which was familiar with the flux of time, knew also those moments when it stood still. The singling out of a moment for the
sake of its experience content or of its significance retarded its
motion or arrested it altogether, so that the phrase *c'est l'heure
où...* produces an effect like that created by focusing a lens on a
point in time. The poetic value of the expression resides in its
highly suggestive working on the hearer or reader. It invites him
to concentrate his attention on a particular point in time, and to
dismiss from his mind everything that lies before or behind it.
C'est l'heure creates an island in time, that is to say, the period
imagined as enclosing the events of the poem. Those events
conceived of as just having taken place seem to recede far into
the past, and the immediate future seems to be separated by a
wide gulf from the present, which appears enlarged as a result
of its isolation. This enlargement is an illusion produced by what
might be called the Romantic temporal perspective.

There are two forms of spatial perspective: the quantitative
on the one hand, which has to do with the relative size of objects
as seen, and the qualitative on the other, based on the optical
impression created by objects; the intensity of this impression
diminishes in proportion as they become more remote. These
two forms of perspective may be said to exist in the temporal
sphere also. The one sets out events, etc., in their temporal rela-

tions and graduates them. This mode of distancing makes use
of the grammatical tenses. These tenses do, it is true, differentiate between periods as regards their position in time, but not as
regards their qualities. The Romantic perspective does, however,
bring out these different qualities. It does not set out periods
in their relative positions in time but serves to emphasize their
qualities—their value as an experience, it may be. By expanding
it and emphasizing its importance, a period is made to stand out
from others.

An example of this particular type of perspective is supplied
by the difference between the use of *jour* and *journée*, a difference
which had gained in sharpness and frequency since the beginning
of the nineteenth century. In speaking of the day which was in
the process of passing or which had just ended, the longer word
was preferred. This is understandable in the light of its original
meaning, 'day's work', 'daily work', 'content of the day'. By
virtue of the co-existence of these two words for 'day' the
Frenchman is able to distinguish the day in general—any day,
an ordinary day—from a particular day which he is actually
living through, and which he sees before him in all its fullness
as an experience. *Le jour* serves to express one's distance from
the day; *la journée*, one's involvement in it. This is a perspective
quite different from that created by 'he comes' on the one hand
and 'he came' on the other.

The mind that was alert to time now discovered all manner of
sensations in the matter of duration. One sort of time appeared
to be diversely refracted, as light in a prism. The retrospective
glance saw in memory the past in a temporally distorted distance.
A phrase such as 'I see him as though it were yesterday', a
frequently recurring formula in the literature of the nineteenth
century, seems to have originated in the sphere of Romantic
memoirs:

> Maintenant que j'écris ceci, il y a dix-huit ans entre moi et ce
> souvenir; cependant, en ce moment même, je vois, aussi distincte-
> ment que si cela s'était passé hier, les traits et l'impression du
> dernier objet qui eut mon dernier regard.
> [Musset, *L'Anglais mangeur d'opium*[19]]

> Voici le lieu de la scène que je vois aussi nettement que si je l'eusse
> quitté il y a huit jours.   [Stendhal, *Vie de H. Brulord*, ch. 13]

K

> Ils sont passés ces jours dont tu dois être fier;
> C'était un autre siècle, et pourtant c'est hier!
>
> [Lamartine, 'Lettre à Alphonse Karr'[20]]

The contrary form of expression places that which has just been experienced in the distance:

> The hours which had just passed lay behind him like long years.
>
> [Novalis, *Heinrich von Ofterdingen*, ch. 5]

> Chaque minute de ma vie se trouve tout à coup séparée de l'autre par un abîme, entre hier et aujourd'hui il y a pour moi une éternité qui m'épouvante.   [Flaubert, *Novembre*[21]]

Our everyday experience of time in our waking state does not reveal its true and innermost nature. Changed conditions of consciousness such as dreaming, a drugged or intoxicated state or ecstasy create a new sensation of time. Nerval tells us of a most unusual and uncanny impression of time he received when in an unhealthy frame of mind:

> Pour moi déjà, le temps de chaque jour semblait augmenté de deux heures; de sorte qu'en me levant aux heures fixées par les horloges de la maison, je ne faisais que me promener dans l'empire des ombres. Les compagnons qui m'entouraient me semblaient alors endormis et pareils aux spectacles de Tartare jusqu'à l'heure où pour moi se levait le soleil. Alors je saluais cet astre par une prière, et ma vie réelle commençait.   [*Aurélia*, Part II, 6]

A person affected by this temporal illusion does not only lose his relationship with natural reality but also finds he has lost contact with his human environment. This is a negative illustration of how a binding temporal order is able to give rise to a state of community.

The 'opium eater' Thomas de Quincey relates how time expanded in dreams in the most alarming way:

> Sometimes I thought I had lived through seventy or a hundred years in one night; I even dreamt of a span of thousands of years: and others which went beyond the limits of anything man can recall.[22]

This strange sensation of time is then ascribed to the waking state also. The temporal boundaries of the self are extended:

> Il me semble quelquefois que j'ai duré pendent des siècles et que mon être renferme les débris de mille existences passées.
>
> [Flaubert, *Novembre*[23]]

A particular moment may acquire an emotional importance and intensity so great that it becomes separated from time altogether. It may appear to the consciousness as completely disconnected from what precedes and follows it. We are reminded of the interpretation placed by the English Romantic, Thomas de Quincey, on the scene in *Macbeth* in which the latter hears a knocking on the door after the murder of Duncan. This knocking brings one to an awareness of the fact that time, which had stood still during the grisly deed, has resumed its normal flow.

> . . . the world of ordinary life is suddenly arrested, laid asleep, tranced, racked into a dread armistice: time must be annihilated, relation to things without abolished.[24]

To one undergoing an extraordinary experience, time seemed to be in abeyance. To say of an experience that it lay outside time was an indication of its superlative profundity.

> un soir, dans la brise embaumée,
> Endormi, comme toi, dans la paix du bonheur,
> Aux célestes accents d'une voix bien-aimée,
> J'ai cru soudain le temps s'arrêter dans mon cœur.
> [Musset, 'Lettre à Lamartine', 1836 (*Poésies Nouvelles*)]

Intensity of feeling might be represented by the poet as complete obliviousness of the outside world and of the time to which it was subject. In the realm of the most intimate spiritual experiences, time—an everyday, commonplace and trite concept—had no place. The Romantics believed they might follow in the footsteps of Rip Van Winkle; it was thought that one could place oneself outside time by plunging oneself into states of ecstasy or into a timeless Absolute, thereby eliminating it:

> Oh! Comme, ensevelis dans leur amour profonde,
> Ils oubliaient le jour, et la vie, et le monde.
> [Musset, *Don Paez*, IV]

This mysticism was a direct product of the new subjectivism. The assumption that a state of joyfulness made a year pass as rapidly as a day may be considered as a comparative, relative to which the idea of the total elimination of time in a state of ecstacy stood as a superlative. Here we have definite evidence of the prevalent notion that time was nothing in itself, and that, indeed, it was experience itself which created and informed time and in certain circumstances neutralized it.

The phrase *c'est l'heure* symbolized yet another view. In it we descry a new mysticism, a belief that certain times were singled out and that they possessed magical qualities. As in the Middle Ages, the hour was held responsible for the events it brought forth, as a person was held responsible for his acts. The idea of a hidden process which went on concurrently with the visible, familiar, investigated and intelligible course of the world was revived. One became aware of the existence of this other world only when its path intersected that of the visible world. Those fateful hours, of decisive importance in the life of an individual, and inexplicable in the light of familiar and visible phenomena, marked such points of intersection. Sainte-Beuve's life of Mme de Pontivy contains the following sentence:

> Un voile couvrait sa voix, un voile couvrait son âme et ses yeux et toutes ses beautés, jusqu'à ce que vînt l'heure.[25]

An unknown span of time had to elapse, the invisible process had to advance to a certain point, the fateful hour had to strike before the mystic event could take place. Time was a mysterious mechanism, an enigmatic signalling system, the functioning of which granted free passage to an event, so far denied it. The wisest course was to have faith in time. It had a better idea than the individual of what was expedient and appropriate. The auspicious moment would surely arrive. The clever thing to do was discreetly to await it. But here we find ourselves again at our starting point.

Notes

1 Xavier de Maistre, *Voyage autour de ma Chambre*, ch. 4 (Internat. Bibl., Berlin, 1922, p. 5).
2 E. de Sénancour, *Obermann* (ed. Michaut, Paris, 1912), letter 9.
3 Consider the function assigned by Fritz Strich (*Deutsche Klassik und Romantik*, Munich, 1922) to the conception of time in his definition of the respective natures of Classical and Romantic individuality; see especially the chapters 'Language' and 'Rhythm and rhyme'.
4 *Corinne, ou, l'Italie*, IV, 3 (Class. Garnier), p. 67.

5 Rousseau and the Romantics were not the first to use *remonter* in this sense, but they applied it to designate the memories of their personal past. See R. Glasser, 'Oben-Unten-Orientierung in der sprachlichen Veranschaulichung der historischen Vergangenheit', *Zeitschr. für rom. Phil.*, 78, pp. 32–58.

6 As will be shown by the author, the retrospective phrase 'I can see' is an old *topos* which gained new favour in Romantic literature of an autobiographical character.

7 Henri Béraud, *Le 14 Juillet. Les Annales* (1 February 1929), p. 101.

8 Book I, ch. 41.

9 A. de Vigny, *Servitude et Grandeur Militaires*, I, 1 (Class. Garnier, p. 14).

10 F. Strich, *op. cit.*, pp. 27, 171.

11 Preface to *Cromwell* (Berlin, 1920), Roman. Texte, p. 31.

12 'Ainsi, tout poète qui aura bien compris l'unité d'action verra dans chaque sujet la mesure de temps et de lieu qui lui est propre.' —'Lettre à M. C.*** sur l'unité de temps et de lieu dans la tragédie', A. Manzoni, *Tragedie e poesie* (Milan, 1873), p. 366. 'Lettre à Lord*** sur la soirée du 24 octobre 1829 et sur un système dramatique', A. de Vigny, *Théâtre complet* (ed. Dorchain, Paris, 1929), I, p. xv.

13 Vigny, *ibid.*, I, p. xvi.

14 *Raphaël*, I.

15 J. J. Ampère, 'Les Renaissances', *Mélanges d'Histoire littéraire* (Paris, 1876), I, p. 440.

16 *Obermann*, 'Observations', IV.

17 *Obermann*, letter 41.

18 Kurt Döhner, *Zeit und Ewigkeit bei Chateaubriand* (Geneva, 1931), p. 123 n.

19 A. de Musset, *Mélanges de Littérature et de critique* (ed. E. Biré, Paris, 1908), p. 19.

20 *Recueillements poét.*, No. 39.

21 Ed. E. Lerch (Munich, 1926), p. 18.

22 Musset, *op. cit.*, p. 93.

23 E. Lerch, *op. cit.*, p. 2.

24 Thomas de Quincey, *Selections* (ed. Dobrée, London, 1965), p. 83.

25 Sainte-Beuve, *Portraits de Femmes* (Paris, 1856), p. 437.

9

Baudelaire's tortured sense of time

In *Les Fleurs du Mal* the subjective view of time of the Romantic period and the later eighteenth century, a view which contained the seeds of later developments, took on an exacerbated form. The follower of the Symbolist cult represented himself as one whose general attitude and will-power could not withstand the current of time, and as one who had neither the desire nor the will to shape time. The only element of his being which he exposed to it was his sensibility, which had reached a state of morbid hypersensitivity. The subjective quality of the days, either dragging or swiftly moving, was exaggeratedly set forth. The poet felt trapped between the Scylla of tedium and the Charybdis of his sense of the transitoriness of things. The basic theme was introduced as early as the beginning of the century:

La vaine succession de ces heures si longues et si fugitives.

[*Obermann*, letter 1]

He was afraid of the time that lay ahead, feeling that his powers would not measure up to it: it set a task, an elementary obligation that could not be evaded. Its coming and going was feared in like measure. To a person in a fundamentally cheerless and bored frame of mind, time seemed to be altogether colourless and shapeless. At best it resembled a barely moving river of mud, of a uniform colour, that crept sluggishly past; its endlessness called to mind a fearful image of eternity. The slowness of time was a symbol of terror which constantly recurs in Baudelaire's work. The words 'slow' and 'slowly' became tokens of his tortured view of time:

Durant ces longues nuits d'où le somme est banni.

[*Remords posthume*]

J'ai souvent évoqué cette lune enchantée,
Ce silence et cette longueur.   [*Confession*]

Rien n'égale en longueur les boiteuses journées.
                                        [*Spleen* ('J'ai plus de...')]

... l'esprit gémissant en proie aux longs ennuis.
                                        [*Spleen* ('Quand le ciel...')]

Je sentis, à l'aspect de tes membres flottants,
Comme un vomissement, remonter vers mes dents
Le long fleuve de fiel des douleurs anciennes.
                                        [*Un voyage à Cythère*]

Et de longs corbillards, sans tambours ni musique
Défilent lentement dans mon âme.   [*Spleen* ('Quand le ciel...')]

                                        le ciel
Se ferme lentement comme une grande alcôve.
                                        [*Le Crépuscule du Soir*]

Tant l'écheveau du temps lentement se dévide!
                                        [*De profundis clamavi*]

Indeed, 'slow' and 'slowly' sometimes became synonymous
with 'ugly' and 'repulsive'. This state of weariness and disgust
knew no regular changes or divisions in time, but only the unpleasant distorted image of it. The days were hideously long.
They limped. Time could be put to no purpose; it contained no
future potentialities which might develop to maturity. This
dreary prospect of the future effectively stifled all hope. Hope
was a bat which collided with cobwebby ceilings.[1] The past was
as dead as the future. Memory was like a graveyard. No matter
whether one looked forward or backward, one saw only nothingness. Sénancour had reproduced this feeling with great accuracy:

Mes jours perdus s'entassent derrière moi; ils remplissent l'espace
vague de leurs ombres sans couleur; ils amoncèlent leurs squelettes
atténués; c'est le ténébreux simulacre d'un moment funèbre. Et
si mon regard inquiet se détourne et cherche à se reposer sur la
chaîne, jadis plus heureuse, des jours que prépare l'avenir, il se
trouve que leurs formes pleines et leurs brillantes images ont
beaucoup perdu. Leurs couleurs pâlissent: cet espace voilé qui les
embellissait d'une grâce céleste dans la magie de l'incertitude,
découvre maintenant à nu leurs fantômes arides et chagrins.
                                        [*Obermann*, letter 46]

Those things normally seen as the concomitants of time—
variety, the difference between the aspect of the past and that of
the future, the expectation of something new—all these were
absent from the sick mind. For it, time did not exist. It saw itself
in a timeless and therefore terrifying world, for the individual
needs time as he does air. The vague disquiet of the pre-Romantics had taken on the intensity of torture. The turn of the
Romantic imagination and forms of thought had already prepared the ground for this. Alfred de Vigny portrayed the situation of a man deprived of time in this way in his poem 'La
Prison', where the prisoner in the iron mask is unaware of time,
but experiences only fearful dreariness instead:

> Pourquoi venir fouiller dans ma mémoire vide,
> Où, stérile de jours, le temps dort effacé?
> Je n'eus point d'avenir et n'ai point de passé;
> J'ai tenté d'en avoir; dans mes longues journées,
> Je traçais sur les murs mes lugubres années;
> Mais je ne pus les suivre en leur douloureux cours.[2]

The picture of a man sentenced to life imprisonment—by no
means a new picture—here contains something fresh: the rack
of torture is tightened by another agonizing notch. The idea of
the prisoner being tormented by time was an old one; time
seemed to him an eternity. But the notion of the sameness of the
past and the future was an innovation created by Romantic
thought. With Vigny, it was an external circumstance that was
the instrument of torture; with Baudelaire, it was the inherent
inability of the individual to experience and inform time, bearing in mind the Romantic proviso that the individual must first
create it.

Time was immeasurably vast and the individual correspond-
ingly small. It swallowed him up as snow envelops a frozen
person.[3] Here we discern a strange contradiction in Baudelaire's
view of time. On the one hand there was time, something
sluggish, dreary and interminable, and on the other there was
the clock, which kept the individual constantly on edge by its
ticking and striking; time passed and he felt it doing so, unable to
utilize and inform it. The contrasting ideas of 'too slow' and 'too
rapid' movement were sharpened in imagination to a torturing
intensity. The clock reminded the person who was incapable of

action that the time, which seemed to him to pass so slowly, had
nevertheless suddenly expired:

> L'horloge, à son tour, dit à voix basse: 'il est mûr
> Le damné!'   [*L'Imprévu*]

> La pendule, sonnant minuit,
> Ironiquement nous engage
> A nous rappeler quel usage
> Nous fîmes du jour qui s'enfuit.   [*L'Examen de minuit*]

> Te convulsant quand l'heure tinte.   [*Madrigal triste*]

> La pendule aux accents funèbres
> Sonnait brutalement midi.   [*Rêve parisien*]

Time was a stupid, unfeeling, indifferent and barely conscious
monster, yet the clock was a highly alert and malevolently diabolical creature, 'l'ennemi vigilant et funeste'.[4] Between time
and humanity there could be no equal combat. Man was the
defenceless victim. The clock, as a symbol of evil, must take
its place amongst those flowers which injected their poison into
human life. In the poem *L'Horloge* it ominously warned one to
take thought of death. Its metallic, impersonal quality was more
horrifying than any power thought of as personal; the rapid
motion and insect-like chatter of the second hand was more
awesome to the ears than the resonance of bells. This was a
frisson nouveau in the history of man's sense of time. This
diabolical grotesqueness of time. which must needs drive men
mad, was a new bogy to join the many others which had reared
their heads. The rhythm of the clock and of other objects and
phenomena had a threatening aspect; their monotonous sound
resembled the footsteps of the spectre of death as it approached
with an assured and inexorable tread. The fear of death now
found the same forms of expression and drew its sustenance from
the same images as in the Middle Ages: those of the skeleton,
the hour-glass and the decaying corpse. But whereas the Middle
Ages represented these things primarily as visual impressions,
the Symbolist, the successor to the Romantic, dreaded also the
sound made by time as it slipped by. Romanticism had made the
ear receptive to the softly monotonous passage of time.

The split and incoherent quality of the Decadent conception
of time manifested itself in the contemplation of death also. The

fear of death was the counterpart to tedium, in suffering which one desired one's own death, imagined as an exciting event; but even this was a disappointment. The future was empty even after death. There were no surprises even in the Beyond:

J'étais mort sans surprise, et la terrible aurore
M'enveloppait.—Eh quoi! N'est-ce donc que cela!
La toile était levée et j'attendais encore.   [*Le Rêve d'un Curieux*]

A careful and thorough evaluation of the role of time in the life and work of our poet, as has been carried out by Peter M. Schon in 'Das Zeiterlebnis Baudelaires'[5] discerns the primordial psychological factor at the root of his poetic utterances: this factor is Baudelaire's detestation of time, born of his helplessness with regard to it, a helplessness which manifested itself to him in the inexorable movement of the clock. Schon makes it clear that this sensation of time and the need to overcome it pervades the poet's entire work. He recognizes the relationship to time in the actual fabric of the *Fleurs du Mal*: 'Baudelaire's experience of time is of such fundamental importance for the interpretation of this volume of poetry that it may be described as the key to the understanding of the work.'[6]

Notes

1 *Spleen* ('Quand le ciel bas et lourd').
2 Imprisonment is a situation without time:

> There were no stars—no earth—no time—
> No check—no change—no good—no crime.

Byron, The Prisoner of Chillon

3 *Le Goût du néant.*
4 *Le Voyage* ('A Maxime du Camp'), st. VII.
5 *Studia Romanica. Gedenkschrift für Eugen Lerch*, ed. Bruneau-Schon (Stuttgart, 1955), pp. 353–83.
6 Schon, *op. cit.*, p. 383.

10
The mastery over time in the nineteenth century

The strivings of the modern world are directed towards the external control of life in space and time. Both appear primarily as media of orientation and power. Time is thereby spatialized and deprived of its value as such to an extent hitherto unknown.[1] Science and technology make of it a means of external measurement and co-ordination. The temporal horizon of historical thought advances into an ever more remote past and the mind becomes increasingly habituated to thinking in terms of centuries and millenia. Past time becomes illuminated and surveyable, for humanity has settled in all the valleys and on all the peaks of the past and has gained familiarity with the landscapes of time. The number of blank places, representing unexplored areas, is constantly dwindling on the temporal map. As historical research serves the past, so does technology serve the present. If the past be a field of inquiry, then the present is a material to be shaped, whether for good or ill.

It would seem that time is one of the most priceless treasures of humanity. But in reality it fulfils a practical function. It is divided and husbanded, not for its own sake, but because it spells wealth and power. No square yard of time in our possession must be neglected or allowed to lie fallow: this is the goal we now seek to attain by a variety of means. The Taylorian system, the stop-watch, the speedometer, all modes of rapid transport and communication, are directed to the same end: to avoid loss of time. The emotional motive of this striving is a thirst for knowledge, mastery and pleasure. It is desired to prolong life by speeding up its tempo, so increasing the number of possible experiences. The more niggardly one becomes with time, however, the more does one become alienated from true duration.

Time is seized and exploited; humanity, ever eager to put it to use, leaves it, as it were, no rest either by day or night. It is for humanity a means to the appreciation of reality. What humanity is able to measure, count, store, catalogue, register, utilize, prove, name and define is for it real. Reality is something that exists, or did exist, on the map or in the universe—something that is accepted as a law and a determining factor in human spiritual life, and which takes its place as a concept in the scientific system. Time has become a certain, surveyable possession. Science relates to measurable time all processes and phenomena to be investigated. Historians are at pains to weave the temporal net of their knowledge of the past ever more closely and accurately. Time, as a divisible entity, also claims their attention in determining periods in historical process.

The human conception of time itself becomes an object of historical investigation. This sets the 'timelessness' of primitive peoples against the temporal order created by the advance of civilization. The temporal element in grammar and vocabulary offers an interesting field of observation to the philologist. In the field of letters the investigation of the temporal structuralization of literary works of art leads to new insights into the processes of poetic creation and style. Attempts are made also, by means of a species of attentive exploration in depth consisting of a 'listening in' to human and poetic utterances, to establish the unique and individual awareness of time of a particular author, and thus to open the door on his general personality. This method is used in masterly fashion by Georges Poulet. In his series of portraits *sub specie temporis*[2] he succeeds, thanks to a finely developed sympathetic understanding, in dissecting the intimate feeling for time of the great intellectual figures of latter times and in isolating differentiated elements of consciousness.

The belief in the relativity of all absolute values and in their refraction by their temporal position in the course of human history has given rise to historicism, for which all human activity appears as a stage of development conditioned by time. The historian's vision ranges far beyond the sphere of human history. The modern 'clocks' devised by science permit of the extensive dating of infinitely remote periods in the history of the world and the universe. The horizon is broadened for humanity by the vision of a temporal continuum which embraces

organic and inorganic natural phenomena alike. Biology in-
vestigates the temporal requirements for the course of vital
processes, for biological rhythms and cycles, and by means of
experiments gains an insight into the perception of time of the
lower animals, so different from the human, and into their
temporal memory. In the light of these investigations into the
small world on the one hand and astronomical infinitude on the
other, relativity lays bare the human measure of time. Psycholo-
gists and medical men distinguish differing 'clocks' within the
human sphere. The commonplace notion that time increasingly
shrinks in the perceptions of a human being during his lifetime is ascribed by the physiologist to changes in his tissues and
a slowing down of chemical processes in his body.

Humanity has gained mastery over time, and not only in the
sense that—to look at the matter in its broadest and narrowest
aspects—it is familiar with every age and turns every minute to
account. It is possible also to perceive and experience the rhythm
and velocity of change and movement otherwise than in their
natural form. Films showing the flight of a bullet or the growth
of a plant may be projected in slow or accelerated motion respectively. We are not, however, concerned with the technical, but
with the spiritual aspects of these developments. This spiritual
aspect is bound up with the emancipation of our experience of
velocity from the physical factors controlling the latter. Humanity
has acquired a weapon against time enabling it to extend the
limits of its experience in time, as the microscope and telescope
enlarge its spatial horizon. It might be possible, with the aid of
technical apparatus, to create temporal illusions which would
cause the relative velocities of familiar phenomena to appear
reversed. The subjectivism of the Renaissance, to which hours
seemed like days and days like years, becomes invested with a
new, objective significance by technical developments in cinema-
tography, by which the events of a second are expanded into an
hour.

The literature of the nineteenth century reflected this objective
view of time in many different ways. The Realist and Naturalist
movements, which aimed at showing life as it is lived, in fact tell
us more of their interpretations of reality. Their reaction against
Romanticism led them to stage an offensive against the Romantic

view of time in the name of objectivity and science. The sanctity of time, veneration of the past and its effects, the inexplorable quality of the future—all typically Romantic notions—were swept away. The new school abhorred *l'imprévu* because it did not represent a slice of life, but only something purely literary. A person familiar with the laws controlling events could not be surprised; he could foresee the future as he could predict the result of a scientific experiment, for the future was composed of the same elements of happening as the past and the present. In the story of the Rougon-Macquart family the end is already contained in the beginning. The new *genre* of the detective novel advanced further along the same path. The cause of an event occurring at a particular time was treated as an unknown quantity x, calculated from other known quantities. Situations were seen as mathematical problems.

Efforts in the practical sphere were concerned with circumscribing the future and reducing it to the level of the present, and with predicting it by synthesizing it on the basis of numerous 'lines of development'—in short, it was desired to eliminate its irrationality. These strivings, the path of which had been prepared by a centuries-old civilization, now reached their culmination. Risk was excluded, death lost its terrors, statistics clothed the menacing and the incomprehensible in indifferent language, insurance companies calmed their clients' fears of the contingencies that might befall them, so that the future did in fact take on the quality of the present, albeit in a somewhat vague and colourless way. Time was spatialized in order to satisfy the general need for security. Future possibilities were directed into a restricted number of channels. This conception of things, which determined the future both as regards time and space with the greatest possible exactitude, might be symbolized as a railway system and a timetable. Even unpredictable natural phenomena had to follow a predestined course. Lightning was directed innocuously to the ground by the human invention of the lightning-conductor.

The future played yet another part in human consciousness. Not only were attempts made to bring it under control and to render it harmless, but in addition a new faith in future ages and generations was born. The present, i.e. the nineteenth century, appeared merely as a preparatory stage which would pave the way for better things; humanity cheerfully and patiently trusted in that future which would be the heritage of posterity. In Renan's *L'Avenir de la Science* the contrast between the real and the ideal took the form of a temporal succession from the present to the future. Accordingly, he expressed his most ardent wishes in the future tense:

> Oui, il viendra un jour où l'humanité ne croira plus, mais où elle saura; un jour où elle saura le monde métaphysique et moral, comme elle sait déjà le monde physique; un jour où le gouvernement de l'humanité ne sera plus livré au hasard et à l'intrigue.[3]

Thus the scientist as well as the practical person was orientated towards the future. So, indeed, was the historian. It would be false to believe that he was essentially orientated towards the past. However great his preoccupation with things past might be, he was concerned, no less than others, with the future. That which he sought and found today was a brick to be incorporated in the ideal future structure of his knowledge. If the future did not exist, his work would be meaningless. The past was only a field of inquiry, not the goal of his efforts. In this he was unlike the dreamer and the Romantic, who turned away from the present and hoped for nothing from the time to come.

The idea of progress, the basis of faith in the future, became such an important element of the general consciousness that it found expression in everyday forms of thought. The belief that what was to come would be better than what existed now was natural in an age of technical invention in which the old was constantly being 'outmoded' by something new and better. This led to a general subservience to time. 'Time' now meant the present, for this was time *par excellence*, the most valuable and important time there was. In order to control it one must subordinate oneself to it. One must adapt oneself to it; the watchwords now were: to understand the signs of the times, not to be behind the times, to move with the times, to keep pace with the times, to be master of one's time, *être de son temps*, to be up to date, to know what time it was, to live with the times, and, if possible, to keep abreast of the times. It seemed as though life was simply a matter of keeping up with the march of time and progress. This attitude was, nevertheless, seen as a serious threat to the self-contained composure of the individual and to the equilibrium of his life and personal development.

The antithesis between time and eternity, which was ever-present in the medieval mind, lived on as an antithesis of values. By virtue of its emancipation from religious matters, it could now be applied to all departments of life. Words and phrases such as 'timeless', 'beyond time' and 'tied to one's time' became frequent. To ask oneself whether something or somebody was within or outside time became a popular method of categorizing historical personalities and circumstances. The mode of evaluating both these alternatives had shifted from its medieval position. Since it was desired to give eternity, as well as the day, its due, the judgment that someone was outside time might mean praise or reproach. This gave rise to an awareness that time placed demands on humanity.

However, humanity now placed its own demands on time. In the fourth and fifth articles of the Declaration of Human Rights of 27 August 1789 the concept of personal freedom was defined in such a way as to lend it an implicit association with the time which everyone had at his disposal. Everybody might make such use of his time as he thought fit, within the limits determined by himself. In this age, when human rights in every conceivable sphere were 'in the air', the concept of the right to time was added to the others. Since, however, the phrase *les droits* had predominantly political associations in the minds of Frenchmen of the epochs of the Enlightenment and the Revolution, we have been unable to find any explicit formulation in the eighteenth century of this right to time. It was not until the following century that the self-evident conclusion was drawn from the general demands of the preceding era for freedom and equality, for the right to time was implicit in these demands. One social demand was for the reduction of working hours to a level consistent with the ability to lead a life worthy of a human being.

Undue concern with the question of saving time in some activity or other gives rise to a new relationship between a person and his occupation. He no longer devotes to it all the attention and affection it deserves. His involvement in a task is in many ways retarded, cooled and constrained by the fact that it makes demands on his time. Where the preoccupation with saving time is uppermost, the idea of steady development and ripening has no place. In no century was there so much talk of growth and development as in this, but in no age was the course of things so forcibly pushed forward as in the nineteenth century. There was talk of organisms, and yet time, which many organisms needed if they were to develop as they should, was fragmented and disfigured.

Thinking in terms of temporal relations in all departments of life—a mental attitude which was only one aspect of a wide-ranging 'mathematization' of the contemplation of reality—penetrated and modified ancient legal concepts also. The most striking change was probably that which took place in the penal code. Offences and crimes were now punished by a deprivation of personal liberty on a temporal basis which corresponded broadly with their gravity. The length of the term of imprisonment enables one to estimate the seriousness of the offence in the view of the judge. Time as a measure provided a welcome means of graduating the penal scale.

Notes

1 The nineteenth century tended generally towards a *déclassement* of time, in the philosophical sphere also: W. Gent, *op. cit.*, p. 91.
2 *Études sur le temps humain*, Paris, 1950; *La Distance intérieure*, 1952; *Le Point de départ*, 1964.
3 E. Renan, *L'Avenir de la Science. Pensées de 1848* (Paris, Calmann–Lévy), p. 91.

11
The rehabilitation of time

The prizing of time as a means to power, and the consequent high degree of importance attached to speed, found a critic in the person of Alfred de Vigny, who, with his sharpened sense for different mental attitudes, saw the advent of the railway as a symptom of the new, interested view of time which dominated the century; from this view he dissociated himself as he called it by name:

> Mais il faut triompher du temps et de l'espace,
> Arriver our mourir. Les marchands sont jaloux.
> L'or pleut sous les charbons de la vapeur qui passe.
> *[La Maison du berger*, I, v. 13]

Haste was a diabolical creature whose bidding men seemed disposed to follow, without knowing whither it would lead them. The poet felt bound to express his disapproval of this mania for speed in the name of poetry and humanity. The idea of 'punctuality' seemed to cast a blight over the world. Whereas formerly the changing landscape, the alternations of effort and relaxation in travel, and one's expectations and their fulfilment had made a true experience of time spent in journeying, all this was now replaced by a straight, soulless line of time linking the points of departure and arrival. He considered that man's relationship to time had been distorted by this new form of transport. In his mind the idea of humanity and of a mode of life worthy of humanity was associated with a certain tranquil rhythm of existence, with which human activity must move in sympathy. He who had no time was no longer a human being, for everything demanded to be contemplated at length.[1] Vigny's protestations against the new and improper use of time, as it seemed to him, had in view the same ideal mode of shaping time as had engaged the attention of Rousseau, though in Vigny's case it was actuated by another driving force. The latter saw men not as creatures of nature and pawns of natural processes but as beings whose essential qualities were intellect, culture and reflectiveness. These were the enemies of all forms of haste and bustle. Mankind must adopt a view of time that was consistent with its intellectual and spiritual well-being.

This criticism was based on reflections so pertinent, and on such an acute intuitive insight into the spiritual vacuum created by the modern deification of time, that in due course other protests were raised as it was realized that a superficial view of time would not permit of a deeper understanding of humanity and human life. The more time was divided, the more hotly it was pursued with new technical weapons, the more apparent did it finally become that a quarry was being hunted that was no longer worthy of the name of 'time', and that this term was a misnomer when applied to it. Towards the end of the century, voices were heard that invested it with a new and dignified meaning. As science went on using time as a gauge for the external measurement of processes, and as the lust for power went on treating it rationally as an object to be exploited, the basic question of the true relationship between man and time was ultimately posed.

It was Bergson who most strongly contested the traditional view of time when he pointed out that two forms of time must be distinguished. The tendency to divide, utilize and control time as a means to an end was reviewed and salutarily checked just as it had reached a climax. Science, with its cinematographic method of apprehending movement, had never come to terms with true duration. Physics had always isolated only points of rest and cross-sections in motion, never motion itself. It had never apprehended the ceaselessly flowing current of time, which represents the flux of reality. What science called 'time' was not true succession and duration at all, but their projection into space. Only the latter is measurable, divisible, reversible and homogeneous. The earlier and the later, today and tomorrow, must not be confounded with one another. The spatialized conception of time blurred and levelled down the essential differences between the past and the future with the object of gaining a rational understanding and mastery of the world. It seemed that Providence had destined human intellect for this sole purpose. Intellect established superficial relations between phenomena, without being able to answer any one of the great questions posed by human destiny. Intellect had a practical function which it could fulfil by disregarding time as an irrational background force and treating it only as a scheme for ordering things. Our true states of consciousness are unrelated to space and quantity. Our experience of time is purely qualitative. Its moments are heterogeneous and merge into one another. They form a qualitative, not a numerical, diversity. For this view of time, the future is uncertain. The future cannot be calculated, for our inner life is not a mathematical but an historical matter. That time which we discover within us by 'listening to' ourselves is an acoustic rather than an optical phenomenon.

If one's attention is fixed on the result of a process rather than on the process itself, time and becoming are shut out. It is a matter of indifference whether, for instance, the quantity 'nine' results from adding 6 to 3 or 3 to 6. In the temporal sphere, however, these two modes of summation are sharply distinguished. Identity of result does not blind the historian to the fact that this result may have been arrived at in a variety of ways. The course of events cannot be arrested; the past still exists in its uniqueness and irreversibility. I may again split the quantity nine into 6 and 3, thereby, in the mathematical sense, restoring the *status quo ante*. But this would be the former state only in a mathematical, abstract way. In the reality of duration previous states cannot be re-created. Subtraction, which reduces a sum to its constituent elements, appears to the spatialized view of things, which leaves time out of account, as a process which cancels that of addition, because this view apprehends subtraction as restoring a former state by reversing the process of addition. In time, however, $3 + 6 = 9$ as well as $9 - 6 = 3$ 'move' only in one direction, because time is unidirectional. That turn of mind which is intent on apprehending externalities and on gaining theoretical and practical 'control' is inclined to give priority to the supposed result of a process, which it arbitrarily assumes to be the true result, instead of apprehending this true result as the outcome of a unique and non-recurring process (in the literal sense of forward movement). It singles out certain stages, seen as results, from the flux of happening, because they seem to it to be noteworthy as observable phenomena or as desirable or assessable end-products of a course of development.

The average person, as he seeks to recover lost time, becomes conscious of his loss when he recalls the past: this remembering consists of elements of thought foreign to events and experiences which actually took place earlier, and of associations which 'pour les besoins de raisonnement' have nothing to do with the essential emotional quality of experience. This 'thinking back', which fails to capture the true nature of things, denatures time. We select from reality those things which further our purposes in action. A large part of our experience is lost to our consciousness by virtue of the fact that it is without practical utility. Marcel Proust set himself the task of seizing and describing this practically useless experience. The notions of 'essential' and 'incidental' lost their significance. A hundred impressions intervene between one's first sight of a person and the first words one addresses to him. They are all equally important. Proust wished to depict life indiscriminately. This was nothing new. The Naturalists were intent on thus indifferently representing reality, so that the author was merely a chance observer. Reality, however, and subsequently time, were understood as standing for something else. In choosing, one was ever mindful of the fact that the reality one observed was itself the result of a choice.

In contrast to voluntary recollection, which moves along well worn channels of thought and is the result of an act of will, stands the involuntary, lightning-like (and, when considered from outside, fortuitous) identification of the present moment with one in the past. This moment from the past, therefore, is neither dead nor lost, but its hidden existence within us is illuminated by a sudden flash. Such moments correspond to that time which we truly experienced and which ever remains in our possession. The fact that it came to consciousness spontaneously is proof of its truth. We feel gratified by its attempt to come to light without the aid of our will and to thrust aside the present moment which called it forth. In this momentary battle in our minds the present is the victor. But our satisfaction at 'owning' the past, which is there in depth, persists. The recognition of the irrational nature of temporal experience brings to mind the *durée réelle* of Bergson. Time does not lie on the surface; as we think, we weave a web over it and so hide it.

The sudden illumination of the past had become a motif for the novelists a considerable time before Proust's day. The characters of Flaubert and Zola found themselves suddenly transported from their everyday environment into an earlier moment of their lives, a moment which they experienced with their senses as though it were present. An unknown psychological mechanism released a true memory-image, which allowed people to jump over time:

> Tout ce qu'il y avait dans sa tête de réminiscences, d'idées, s'échappait à la fois, d'un seul bond, comme les mille pièces d'un feu d'artifice.   [Flaubert, *Mme Bovary*, III, 8]

> Peut-être revit-il, en une vision rapide, là-bas, à Moscou, sa maîtresse pendue, ce dernier lien de sa chair coupé.
>   [Zola, *Germinal*, VII, 2]

It seems that Proust was linguistically anticipated by the use, which had become general in narrative literature since Flaubert, of the words *revoir*,[2] *retrouver*[3] and *revivre*[4] to designate the global actualization of the past. It is true that the author of *A la Recherche du temps perdu* attached a particular importance to rediscovery and made it a pivot and constructional principle for his novels.

But it would be quite wrong to regard Proust as a fashioner of mere *mémoire affective*.[5] The bringing to life in memory of isolated childhood impressions is, for one who expects no confirmation of his existence from the future, a vital process, and one which his intuition assists in intensively realizing. The past gives the present its reality.[6]

> La seule connaissance de soi possible, c'est donc la re-connaissance. Lorsque à l'appel de la sensation présente surgit la sensation passée, le rapport qui s'établit fonde le moi parce qu'il fonde la connaissance du moi. L'être que l'on reconnaît avoir vécu devient le fondement de celui que l'on veut vivre. L'être véritable, l'être essentiel, c'est celui qu'on reconnaît, non *dans* le passé, non *dans* le présent, mais dans le rapport qui lie le passé et le présent, c'est-à-dire *entre les deux*.[7]

But these moments, in whose rebirth the poet sees his self confirmed, remain mutually unrelated points of light which do not form the continuum of a true duration. The remembered states of consciousness, standing in juxtaposition like the parts of space, and thereby distancing the Proustian conception of time from the Bergsonian, demand to be synthesized in the mind of the one evoking them, who must link their general, timeless relationship with his own existence and thereby rediscover lost time and lost space—for, in the Proustian mode of remembering, scenes as well as moments are isolated; they resemble unconnected vessels. Whereas Bergson attached the highest importance to separating time and space sharply from each other, Proust's novel of time, in the final analysis, comes to reducing the temporal succession of states of consciousness to their spatially ordered co-existence. He opposed the Bergsonian flowing and becoming with a search culminating in the acquisition of mental possessions.

> Ce qui est beau à Guermantes, c'est que les siècles qui ne sont plus essayent d'être encore; le temps y a pris la forme de l'espace mais on le reconnaît bien,

he observed in *Contre Sainte-Beuve*.[8]

Despite the widely divergent results of their respective views of time, Bergson and Proust each in their way contributed to a revaluation of time in the consciousness of the twentieth century. Their views may be understood as mutually independent attempts to overcome, in the philosophical and aesthetic domains, a strongly marked positivism in thought and life.[9] In this respect their forerunners, aiming at the same goal, were the Renaissance and the Romantic movement. Their effect in breadth and depth on European thought is inestimable, the more so as the Proustian premise of the experiencing of vacuity and disconnected moments seems to be a widespread feature of the contemporary poetic world view.

Whereas on the one hand the modern person's feeling for time is alert and has become widely differentiated in keeping with his developed individuality and as the heritage of a rich cultural tradition, on the other the secular process of the exploitation of time as a quantity goes inexorably on. Its result is the widespread parcelling-out of the time at the disposal of humanity, as individuals and groups, as a condition of social progress and the ordering of communal life in its thousand ramifications and its temporal constraint. What a contrast with former centuries!

> Les hommes n'avaient pas encore été contraints par les rudes disciplines horaires que nous connaissons: l'heure civile, l'heure religieuse, l'heure scolaire, l'heure militaire, l'heure usinière, l'heure ferroviaire.[10]

In fulfilling this function the objective and universally binding temporal order is something of the highest importance and value, an acquisition from whose advantages even the extreme proponents of unrestricted temporal freedom for the individual draw profit. The European is so well adapted to the temporal order he has himself created that he is inclined to appreciate it as a significant regulating force in public and private life rather than to groan under the tyranny of the calendar and the clock.

Notes

1 Car il faut que ses yeux sur chaque objet visible
   Versent un long regard.
                              La Maison du berger, I, st. 19.

2 'Elle revoyait Venise et leur séjour dans cette ville... Elle revoyait les mois qui avaient suivi le retour à Rome... Elle revoyait l'épisode de leur rencontre...'—Paul Bourget, *Cosmopolis*, VI (Paris, 1893), p. 196.

3 '...il retrouvait, tout le long de la petite écriture qui lui disait des phrases si douces, les émotions oubliées d'autrefois.'—Maupassant, *Fort comme la Mort*, II, 5 (Paris, 1896), pp. 280–1.

4 'Il revécut son amitié pour Marcel...'—Henry Bordeaux, *La Peur de vivre*, II, 6 (Paris, 1902), p. 100.

5 G. Poulet, *Études sur le Temps humain*, p. 374.

6 '...phénomène d'ordre plus métaphysique que psychologique'. Louis Truffaut, *Introduction à Proust* (Munich, 1967), p. 119.

7 Poulet, *op. cit.*, p. 393.

8 *Contre Sainte-Beuve* (Paris, 1954), p. 285.

9 Cf. Hans Robert Jauss, *Zeit und Erinnerung in Marcel Prousts 'A la Recherche du temps perdu'* (Heidelberg, 1955), p. 13.

10 L. Febvre, *Le Problème de l'incroyance*, p. 430.

Index

Persons

General